**Twopence** ...

by the same author

THE DEVIL'S PIPER

# Twopence a Tub

*Susan Price*

FABER AND FABER LTD
3 Queen Square, London

*First published in 1975*
*by Faber and Faber Limited*
*3 Queen Square London WC1*
*Printed in Great Britain by*
*The Bowering Press Limited*
*All rights reserved*

*ISBN 0 571 10624 2*

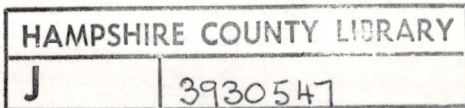

# 1

Two young men walked shoulder to shoulder around the worn earth path at the edge of the quarry, through the dusty long grass and nettles. It was a late evening in summer and the light was heavy, as was the stink of pig-sties, muck-heaps and factories. It was Sunday, and church-bells jangled dreamily, near and far away.

One of the two men was tall and thin, lanky. His arms and legs seemed to be fastened with pivots rather than joints, so that they swung about in unexpected directions. He was dark, long-faced, long-nosed, with thick black hair bunching from under his cap, sticking out at collar and ears. His skin was a washed-out yellow, and this gave him an unhealthy look. His name was Thomas Shannon, and he was the eldest son of Thomas Shannon; so he was called Shanny by everyone, to distinguish him from his father.

He was a collier. This could be seen from a distance because he walked very upright, as colliers do above ground, and because he swung his arms across his body, elbows out—a habit colliers fall into underground, where there isn't much room. Close to him you could see the blue tattoos on his face, lines and dots across his nose, on his brow and cheeks. The tattoos weren't made with needles and ink; they were scars left by pieces of falling coal and stone. The skin had healed over coal-dust in the wound.

The other man was shorter and neater. He walked quickly, in a straight line, not wandering about as Shanny did. His face was squarer than Shanny's and more likely to turn sullen. No hair showed beneath his cap, except for the fore-lock, and his skin was as white as a baby's, through never having been very long in the sun. His name was Jechonias Davies, and he was the eldest son of Dewi Davies. He was called Jek, because that was easier to shout than Jechonias.

Shanny and Jek were cousins, their mothers being sisters, and they lived in houses that backed on to each other, so, in all their lives, they had never been out of hearing of each other. Shanny had been born in the last week of December, 1840, Jek in the first week of January, 1841, making Shanny almost exactly three days older than Jek. Then they had been brought up together, hardly able to tell their mothers apart for the first five years; sharing food, clothes and toys; playing truant from school together; and going down the pit on the same day, April 2nd, 1851.

They left the earthen path and lurched off, shoulder to shoulder, into the long yellow grass and nettles, the grass-heads rattling on their trousers. They were tired, and showed it, sometimes swaying together, sometimes stagger-ing apart. It had been a long day.

Not far off dark when they'd set out that morning, two o'clock it must have been, deep blue light, and cool. They walked to Dudley across the fields, a distance of little over a mile, so it couldn't have been later than half-two when they reached the town, probably earlier. At that early hour you'd think a town would be dead, but not Dudley. The streets were full of people, most of them drunk, laughing, singing and arguing. A sinful place was Dudley, so Jek had been told, not a place for young lads, especially after dark when it wasn't safe to be alive among the dark alleys and back-streets. If you were dead they'd leave you be, but alive they'd smash your head in for a penny and your jacket. Nevertheless, they dawdled through the streets, and excepting a drunken man who fell against them, threatening to thrash them if they didn't give him a twist of tobacco, no one so much as shouted 'Boo!' The drunken man had given them a scare, but then he'd fallen down and they'd stepped over him and gone on, admiring the night-life. Until two ladies had taken their arms and breathed stale beer and onions over them. Then, flustered, they'd fairly run out of the town to the country beyond it, away from the stinking, crowded, black and noisy little streets.

Jek didn't like the country very much; it was all too strange for him. He didn't like the funny smell, which was fresh air, nor the empty spaces, nor the quietness and the quiet sounds. Bird-song and wind in tree-branches weren't for him. He preferred factory hooters and drop-hammers. Shanny knew more of the country and was happy there—Jek followed him in silence.

It was still early, about seven, when they climbed the wall of the estate—carefully, because of the glass embedded in the top—and dropped down into an overgrown, neglected corner. With all the land the lord had fenced in and rented out, he could afford to ignore parts of it. But if he'd known the poachers that used it were so many, he'd have had it cleared in no time.

By now Jek was feeling exhilarated; perhaps it was the unaccustomed amounts of oxygen he was breathing in. But the silence and the strangeness kept him obedient to Shanny. Shanny told him to be careful of man-traps, to watch where he was putting his feet, because if Jek got caught in a man-trap—well—he'd wish he hadn't, that was for sure.

They visited all the nearby Shannon snares, and took seven rabbits. Shanny moved quietly, though he made no pantomime about it. He said that if they moved quietly and steadily they wouldn't frighten the birds much, and the birds would warn them of another person being near by

9

giving their alarm call. As they moved into the trees, a flock of wood-pigeons rose up through the branches. Good, Shanny said, that meant that no one had been here for some time before them.

Then back over the wall and as far away as possible as quickly as possible. They didn't want to be transported.

At a safe distance they stopped, and Shanny skinned and gutted the rabbits with his clasp-knife. Jek stood by, kicking the heads into the bushes, watching Shanny cut off all the paws and put them in his pocket, to sell for good-luck charms. The carcasses he dropped into the big poacher's pockets he had stitched inside his jacket. They walked on to the nearest village, Kinver.

It was noon when they arrived, through having wandered from the way and climbed trees, and they knocked at the back door of the pub, bringing the landlord from his dinner. Jek sat on the kitchen doorstep, in the yard's sun, and waited sleepily while Shanny bargained somewhere in the pub's cool innards. The landlord's wife gave him a cheese sandwich and a gill of beer.

Shanny came out, at last, with ten lovely shillings and three rabbits that the landlord had refused to buy. Shanny said to hell with it, they wouldn't try to sell the last three. The Shannons would keep two and the Davieses could have the other. He gave Jek two shillings, carefully counting out all the coppers, and keeping the sixpences and silver threepenny-bits because they were bright and he liked them.

Jek thought Shanny very generous, foolishly so. After all, he'd only come along to keep company and he'd done none of the work. He thought guiltily that if he'd been the one with ten shillings, he would have given Shanny only sixpence or a shilling, for appearance's sake.

After they'd rested and begged another gill of beer from the landlord's wife, they started home. It was afternoon and hot, so they stripped to the waist and were still red-faced and damp. Jek didn't like the sun in the country. It was too harsh without its screen of smoke.

They didn't go straight home, but cut across Oakham way to the Blue Fly Pit, well into the old, comfortable stink and smoke. The Blue Fly colliers had gone on strike a week ago, a very exciting thing, and they wanted to visit them and take back home news of the strike and various relatives to people at their pit. The Blue Fly received them well and they sat on doorsteps drinking home-brewed beer, admired babies, pigeons, dogs and fighting cocks, listened to talk of the strike and learned messages—'from me to Our Kid—mind it.' Nothing had been heard from the Blue Fly Gaffers, but their Union Man was meeting with the Union men from the Fair Lady and Ramrod Pits, to see whether those Pits could stop work too. That would be more power to the Blue Fly's arm.

Now they came home, at nearly ten o'clock. Jek would have been shouting angry if he hadn't been so tired, and so hungry; and with the prospect of getting up at half-past three the next morning for a heavy day's work hanging over him. Instead he felt limp and dispirited. "Why don't we come out on strike for the Blue Fly?" he said. "Why don't we, eh?"

"Not enough money," Shanny said.

"Ar! Not enough money. That's always the excuse. They've been collecting our sixpences for long enough, ain't they? They ought to have enough by now. Instead of keep *talking* about a strike, why don't they go on strike?"

Shanny shrugged and grinned vacantly. His teeth were large and yellowed; a blue scar ran across his lips, and there the teeth were broken, the upper tooth barely showing above the gum, the lower teeth filed to pointed stumps, like a cannibal's. "We're all right, Jek, we're fine. We don't need no strike."

"We're all right?" Jek cried. "We ain't all right!"

"Why ain't we?" Shanny asked. "Thee got a dinner waitin' for thee, ain't? Thee got clothes on tha back, ain't?"

They came to the edge of the field, where it sloped down to their home. Jek grabbed Shanny's arm and pointed dramatically. "*That's* why we ain't all right," he said.

Shanny looked down on the place where he'd been born, and couldn't see what Jek meant. He couldn't see that there was anything wrong with the place. Home, wasn't it? Better than a kick in the teeth, wasn't it?

They looked down on a road; unpaved, of dried earth. On the far side of the road stood the Baden Tool Works, a huge, black building with tiny barred windows and a dark tunnel-gate, with a portcullis, and a clock set over it.

On their side of the road, at the bottom of the field slope, was the White 'Oss Row, where they lived. The Baden Works, and the houses near it were built along the road, but the White 'Oss Row was built at right-angles to it, showing only a blind wall to the street. To reach the doors of the Row, you had to leave the road and walk up a dirt-track. A stream, the Baden Brook, ran down the track in front of the White 'Oss Row. On the other side of the brook was the Baden Tavern, a red-painted pub.

Behind the Row was a brick-walled field, full of pigs, and then open fields, running all the way to Dudley: yellow, rank fields, scruffy like a goat's coat, and pocked with marl-holes and quarries.

The Row's real name was Baden Row, because it was built directly opposite the dark gate of the Baden Tool Works, but it was called White 'Oss Row because it was owned by the White 'Oss Pit. Every man and every boy over ten who lived there worked at the Pit.

There were eleven houses in the Row, but they had all been divided into two with thin partition walls, to make twenty-two homes. The houses tottered; some leaned forward from the line, some leaned back; some had simply slipped down into the ground, and the end house hung over the road at an unbelievable angle. The roofs curved up and down, humped and bowed like a camel's back. Subsidence had caused the twisting: the galleries of the White 'Oss Pit ran under the Row, and the ground was falling into them.

Each home had one window up, one down, neither made to open; and one coffin-wide, five-foot-tall door. There was one room down, with cooking-range; and one bedroom up.

There were no lavatories and no water-supply. The colliers had built themselves a lavatory, called a 'boggin'-hole', at the far end of the Row. It was a rough shack with a piece of sacking hung over the door; inside, a trench, and a plank with a hole cut in it, balanced over the trench on bricks. When the trench was filled with muck the women and children emptied it and piled its contents around the boggin'-hole. Sometimes it would be taken away for fertilizer, often it wasn't and then it decayed into a stinking fluid, ran into the brook and the houses, and into the road.

In winter water could be fetched from a spring that rose in a marshy part of the field a step or two away from the houses. It was fresh, clear water, but the spring was obviously fed by the Baden Brook, because when the brook ran low in summer, the spring dried up. It was June now, and the spring was dry. So for water they had to walk two miles into Oldbury, to the well, which was a muddy hole in the ground; or take the water from the Baden Brook. The brook was polluted with effluent from the boggin'-hole, and from other boggin'-holes and pig-sties which it had passed through on its way to them, but the people still drank the water, because it was easiest to fetch.

In the winter, though, there was more water than they wanted. The spring came back at full strength, so did the brook. Usually it rained heavily; the brook flooded, swirling mud, muck and rubbish into the houses, washing offal from the slaughter-houses about the streets. The track and the streets turned to a thick, knee-deep mud, which wasn't only a mixture of water and dirt; and the houses became as cold and damp as they were sweltering and damp in summer.

"What's up?" Shanny asked, mystified, turning back to Jek. "What thee mean: 'That's why we ain't all right'?"

Jek didn't answer. He was staring with wide eyes that saw no detail, but saw everything there was to see. The brick, soot-covered; the last of the week's smoke, still hovering over Sunday; the filth, the crowded buildings. He thought that there was beauty in it, but only because the ugliness took the breath away, and it was wrong. Wrong, wrong, wrong.

No need to be rich as a king, or beyond the dreams of avarice, as they said in the plays at the Palace of Varieties in Oldbury. Didn't want any houses as big as the Baden Tool Works. Just something better than what they'd been allowed. It wasn't asking much.

He knew that there *was* something better than this because he'd seen it, out in the country. Great white houses with gardens big enough to hold a dozen Rows—all for one family. Hard to believe, but as true as he was standing there.

Jek came back to himself with a jerk, and hurried down the slope to the level of the dirt track. Shanny was waiting for him, kicking at a heap of muck with fluid oozing from its base. Together they walked down the Row, between similar heaps. They called at houses on the way, to deliver the messages.

Jek's house was half-way down the Row, on this side. Shanny's backed on to Jek's house, on the other side, so Shanny left him and went on down the Row, and disappeared behind the hanging house.

The Davieses' house had fallen backwards into the

14

ground, and the two steps up to the door had been cracked away from the wall. You had to climb over them to reach the door.

The door itself was open, because of the summer heat, and it hung into the room from its hinges. As you looked in through the opening the whole room sloped away from you and down. The tiled floor had buckled and cracked, sinking in places. It had sunk in front of the door, so, to get in, you had to jump. There was a worn place on the door-frame where your fingers fitted as you steadied yourself.

To make it more difficult the big heavy wooden settle had been placed inside the doorway, so that the open door caught against it. The door had to be shut before you could squeeze past the settle. Jek shuffled, gripped the door-frame and swung in like a monkey to crash against the settle. "I do that every bleedin' time!" he said. He shut the door to, in order to pass the settle, and then came up against the table, which stood squarely in the centre of the room, surrounded by straight-back chairs. There was just enough room on three sides of the table to edge past, if you turned sideways; across the table, on the hearth, there was a little more room, but not much.

To the left of the table was the window: small, clean, but obscured with yellowed net curtains, a strip hanging limply at each side, a strip across the top and one across the bottom. With all that net, and the window slanting as it did, you got a very queer view of the outside world. As if you were rats peering up from a hole, through thick cobwebs.

At the other side of the table was a rickety flight of stairs, clinging to the buckling division wall. They were steep, the steps narrow and worn. There were no banisters.

Across the table from Jek was the hearth, a narrow oblong of tile in front of the cooking-range, covered by a rag-rug, made by pushing bundles of rags through the weave of an old sack. On the window side of the hearth was

15

a straight-back chair; on the stairs' side a three-legged stool, its seat shaped by much use like a backside.

The cooking-range filled the wall between the window and the stairs, and went back into the wall. There were two hobs, for boiling kettles and saucepans, with a fire always burning in the grate between, winter or summer. One hob was hollow, to make an oven for baking, or stewing meat in jars. At the back of the range, in the recess, was a hinged bar with a hook on the end, the gale-hook. The gale-hook swung out over the fire, and pots or meat could be hung on it.

The mantelpiece that ran along the top of the range was very high. Jek had to lift his arm in order to lean his elbow against it. A frill of faded material ran along the shelf's edge, and a brass rod just below that, for towels to be hung on, to dry. There was no clock—they leaned out of the door and looked at the Baden's clock when they wanted to know the time—but there were two black and white china dogs, one at each end of the mantelpiece. Also a shapeless, gaudy lump of chalk that was supposed to be a Scots Highlander and his Lassie. Jek was fond of the dogs, but he detested the Highlander.

On the back of the whitewashed planking door, coats and belts and caps were hung, bulging into the room. From the ceiling hung bags of potatoes, strings of onions and parsnips, none very big, because they were home-grown. Above the settle was a shelf, loaded with crockery, salt, candles and groceries. Jek's father had put up the shelf, and the wall was cracked where he'd driven in the pegs to hold it. The wall was too weak to hold up a shelf and itself as well.

The walls had no covering but their plaster and in places this was missing. A black crack ran from half-way down the wall to disappear behind the settle; others starred out around the window and from behind the stairs, or ran across the ceiling. Subsidence again.

16

The Gaffers said: The cracks are caused by subsidence. It is of no use repairing the cracks until the houses have finished settling into the ground.

The colliers said: And that won't be ever.

In the room, besides all the furniture and Jek himself, were six other people. On the chair by the hearth sat Jek's father, Dewi Davies, in his shirt-sleeves and working trousers: a short, big-nosed man with a mouth like a line drawn across his face. None of his children took after him in looks, except that they had the same dark colouring. None of his children liked him much, because he found them wanting. He thought that they weren't as brave, or as clever, or as sensible as he had been when he was their age.

Reenee Davies, Jek's mother, sat at the table, head down, arms sprawled over the cloth in front of her. Sometimes, when she was in a pleasant, motherly mood, Jek would look at her and think that she must have been very pretty—eighteen or so years ago—being fair-haired, straight-nosed and blue-eyed. But his birth had put an end to all that; he knew, because she often told him so. The other children, and working to look after them, and Pit-work too, had not improved her looks or her temper, but Jek had been the start: the cause of her heavy body, greasy skin, red face, rough hands, screeching voice and weariness. She blamed him.

Sarah and Aynoch lounged on the settle, Joe on the stairs. Jek had known them all since they were babies, had been washed with them in the Baden Brook, had washed them, and had slept with them—but he didn't know them. He didn't really know any of his family. Perhaps they shut off their thoughts from other people, because they had to have privacy.

Nellie, the baby of the family, sat at the bottom of the stairs. She was four and they were beginning to trust that she would live, and take notice of her. The baby before

Nellie, Thomas, had seemed healthy up to the age of one, and then he had sickened and died, as the boy born before Jek had died. You had to watch yourself with babies and not grow too fond of them, because there was always the chance that they would die.

"Hello, chicken," Jek said to Nellie, and she came crawling to him through the chair-legs. Jek picked her up and held her high against his shoulder, one arm under her soft bottom. She was naked, for coolness, and Jek rubbed his mouth against her soft little shoulder, to feel the softness, and laughed when she put both arms around his head and hugged tight, fit to burst out his brains.

"Put tha shirt on," Dewi said sourly.

"It's too hot, Dad." Jek broke out in a sweat as he began to edge around the table, with Nellie, to the hearth. The whole room was blanketed thickly with heat, even with the door open. Jek sat down cross-legged on the hearth-rug with Nellie in the cup of his legs, her legs across his ankles. He clasped his hands around Nellie's waist and she curled over them like a kitten. He pushed his face into her soft hair and blew down her neck to make her wriggle. She was the only soft thing in the world.

Dewi's fore-finger rapped on his skull. "Where thee been till this time? It's past ten, *I* know. Stop playin' with the babby and answer me when I speak to thee, else I'll give thee summat as'll make tha tongue wag."

"I've only been out to the country, Dad, and to the Blue Fly. With Shanny."

"Been to Dudley, haven't thee? An' what was thee doin' there, eh?"

"We only went through Dudley, Dad. We didn't stop."

"Well, don't. If ever I hear of thee bein' in Dudley—I'll give thee a weltin' thee won't forget in a hurry, hear?"

"I know, Dad, I know," Jek said.

Dewi cuffed him around the head. It was mild for Dewi,

one clout with the open hand—he must be in a good mood, Jek thought as his head snapped sideways and his brain buzzed. "Thee know *nothin'*," Dewi said. "Thee'm just a child as yet, an' thee mind that. Thee get funny with me, me lad, I'll knock it out on thee."

Jek shuffled away from him across the hearth. "Come back here!" Dewi roared. Knowing what was good for him, Jek obeyed, and received a clout to the back of the head that knocked his head forward and cracked it against Nellie's. Nellie began to bawl, so Dewi slapped her mouth and said, "Now I know what thee'm cryin' for, tha little bleeder."

No one except Dewi, Jek and Nellie took any notice of this. Dewi hit people when he was bad-tempered, and Dewi was always bad-tempered to some degree. It was normal.

Reenee suddenly roused herself and said to Jek, "Thee wasn't here for thee dinner again."

Her tone was not pleasant, so Jek didn't look at her, only stared at the fire. "Tha dinner was spoilin', so I let the others eat it," Reenee added, and there was a trace of spite in her voice.

Jek's hunger immediately increased and sharpened. After a pause he said quietly, "Couldn't thee have put it in the oven?"

"Watch how thee speak to tha mother," Dewi said.

"I didn't feel like messin' and modgin' about," Reenee said. "I dished tha dinner up, tha didn't come, so I said, go on eat it and serve him right. Learn him to come home for his meals. I don't see why I should always be working myself to the bone for a pack of ungrateful swine like thee."

Again the pause while Jek struggled not to say what he wanted to say, because of Dewi. "I'm *hungry*. I'm clammin' to death here."

"I ain't goin' to slave for thee," Reenee said. "I know thee. If I'd saved it thee wouldn't have eaten it."

Jek choked. "Wouldn't have eaten it!" But it was useless. There was no dinner and that was that. Tomorrow was going to be a long, hard, hungry day. He would bet it was Aynoch who'd scoffed his dinner—just let him wait.

Five minutes went silently by, the fire cracking and whispering to itself, the sweat wandering down Jek's face and chest. Then Dewi said, "Thee wasn't at Union meetin'."

"Gawd," Jek said. He shook his head. The Union meeting, held every Sunday in the old stable behind the Tavern. What good was it? They never did anything. Only talked and talked about some time holding a strike; some time, never.

"Thee went up the Blue Fly, did thee?" Dewi asked. Jek was puzzled by this sudden change from the sore point of his weekly non-attendance at the Union meeting. "Ar," he agreed cautiously. He struggled with Nellie, who was trying—painful, it was—to kneel on his thighs and blow down his neck.

"How are they gettin' on?" Dewi asked, and there was something wrong about this questioning. It was round-about, and Dewi was never roundabout except when he was playing a trick.

"All right," Jek said. "They reckon the Fair Lady an' the Ramrod are strikin' an' all."

"They're strikin' an' all, are they?" Dewi asked, and sniggered. Jek could hear Aynoch sniggering too, and Sarah giggled until Dewi silenced them both with one powerful glance. Jek was baffled. They were laughing at him, there was some joke in what he had said—and he couldn't see it at all. There was nothing funny about strikes. They were—grand. Noble. Good. Jek was hurt that his family should think a strike funny, when he thought it so serious.

Reenee raised her head again. "If thee ain't goin' to work

20

tomorrow, one on thee can fetch me some water. For the washin'. An' thee, Sarah, can help me. It'll be nice to have some help for a change."

Jek thought he hadn't heard right. He stopped curling Nellie's hair around his fore-finger and twisted round. "Eh? What thee say, Mother?"

"Thee 'eared," Reenee snapped.

"But—ain't going to work tomorrow . . . what thee goin' on about?"

Sarah was giggling again. Aynoch said sleepily, "What thee go an' spoil it for, Mother? What thee always spoil things for?"

"I don't play tha saft games," Reenee said. "Thee ain't going to work tomorrow, Jek, because . . ."

"Shutup, shutup!" Aynoch shouted.

". . . because these saft sods have gone on strike. Just because their friends from up the hill don't know what's good for 'em and go on strike, this lot have got to do the same. So while they're strikin' for more pay, we've all got to starve an' clam. I never heard anythin' so stupid in me life."

"Strike?" Jek said helplessly. "Eh? What's this? When did this happen?"

But no one had any time for his questions. Dewi shouted at Reenee, "Why don't thee keep tha gob shut? Thee'm as ignorant as thee am high, thee've no idea what thee'm talking about, hast? If we get more pay thee'll be there with tha pinny held out, won't thee? Ar!"

"I know that we'll go short for food, but thee won't go short of drink, for all I'm ignorant!" Reenee shouted back. They leaned towards each other across the table, bawling into one another's faces.

"I've always give thee money when thee've asked for it, ain't?" Dewi demanded, half rising.

"No!" Reenee's voice rose and cracked so that Jek

21

gritted his teeth. "No, thee ain't! Thee've only give me money when thee've felt like it, or when I've fetched thee out the pub. If I didn't go and fetch thee out of pub every Friday night that God sends, *I* shouldn't see any of the money at all!"

"That's a lie!" Dewi said, standing up, and Jek snatched Nellie up and ducked aside, to the stairs. He didn't want to be within reach of Reenee or Dewi.

The fight was stopped before the punching began by Grandad Ellis arriving at the door. Sarah jumped up from the settle to help him into the room. He was a very old man by anybody's standards, eighty years he'd lived. It was an almost incredible age for a collier. But in all that time, he'd not learned to keep out of other people's arguments. "What's this, what's this?" He glared at his son-in-law and daughter. "This isn't fit for the childer to hear, this isn't."

Dewi flung out a hand to point at him, and said, "Thee keep out of this. I didn't ask for thy opinion and I don't want to hear it."

The old man looked haughty until Reenee, folding her arms, sniffed and said, "It's this strike, Father."

The old man nodded. "Ar, the strike. Saftness, that's all that is. Thee won't win."

"I said I don't want to hear thy opinion!" Dewi yelled and slapped the table.

"What about the strike?" Jek pleaded. "What's this about a strike?"

"Thee don't want to hear my opinion because thee know it's right," Grandad Ellis said calmly. "Strike's against God and it's against man. It ain't right no way thee look at it, and thee can't win."

Dewi pointed at the old man and said to Reenee, "If I don't throttle him before he's swallowed another crumb o' *my* food . . . I'll swing for him. I mean it, I will if thee don't keep him under."

22

"Father—sit down," Reenee said. With his head held high, Grandad Ellis edged around the table and went to sit down on his stool. Dewi stood by his chair and seethed uselessly, because the argument had dropped and he had nothing to say.

"What's this about a strike?" Jek cried out from the stairs. "I keep askin' an' nobody tells me—what strike? Why ain't we goin' in tomorrow?"

For a minute he thought that he still wouldn't receive an answer. Everyone was looking to Dewi, and Dewi was still angry. But then he suddenly snarled out, "We're on strike, the White 'Oss is on strike! That's what was decided at Union meetin'—that we'd take a chance and strike now, with the Blue Fly and the Fair Lady. We couldn't find everybody to ask 'em if they wanted to strike, but everybody that was at the meetin' voted in favour, and they was enough to carry it. So we're on strike, we'll show 'em!"

All heads turned to Jek, to see what he thought, and to hear what glorious words would fall from his lips now that they were at last on strike as he had always wanted. "Damn it all!" Jek said. "I missed it!"

"That's thy fault," Dewi said.

"On strike!" Jek shouted. "We'll show 'em, won't we? For more pay? How much?"

"We're going to ask for a penny more on every tub filled with coal," Dewi said solemnly.

"Twopence a tub," Jek said. "That's a lot. Hey, but they'll never pay that."

"It's what we're askin' for," Aynoch said, eager to get into the talk. "It ain't what we expect to get. Bargaining like. If we get half-penny more that's—about—four shillin's more for me."

"An'—four an' nine for me," Jek said. "Four an' a tanner for thee—Hey! I wonder if our Shan knows."

*   *   *

The Shannons were glad to see Jek, as they were glad to see almost anybody. They were celebrating the strike with home-brewed beer, whisky and cake because they believed in any excuse for a celebration. Tom Shannon filled Jek a jam-jar with the whisky; Ruby Shannon filled his hand with greasy, curranty bread-pudding.

Jek said, "Ain't it bostin'? We'll show the Gaffers now, we'll settle 'em!" He was surprised and hurt when the Shannons started to laugh. "What's so funny?" he demanded.

"Aah, get away Jek," Tom Shannon said. "We won't win the strike, not a chance—but it's a lovely holiday. Drink it, Jek, drink it, for God's sake! Knock it back like this." He threw back his head and drained his jam-jar. Jek took a large gulp, and found that the whisky was easier to drink that way. It didn't burn until it hit the stomach lining.

"Jek," Shanny asked, sprawling over the floor at Jek's side, "thee sleep in the field tonight with we?"

It was the habit of the Shannons to sleep in the open during hot weather. In this, the downstairs room, they had a table and two crates—one for Thomas Shannon to sit on, and one for Ruby—a stewing jar, two saucepans, a kettle, a pile of plates and spoons and a selection of jam-jars for use as cups. The skirting-boards had been ripped from the cracked walls long ago, for fire-wood; some of the tiles had been pulled up, for purposes unknown, perhaps to make the floor more level; and a hole had been smashed straight through the wall under the window, so that the cat could come in and out without their having to open the door for it.

"No," said Jek. "I don't like sleeping out. Tha get wet."

"Aah," Shanny said scornfully.

"Leave him be," Ruby Shannon said. "There's them that like sleepin' out an' them that don't, and them as do should leave them as don't be. Have some more o' this puddin', Jek, afore they eat it all."

"It's plain that we can't win," Tom Shannon said, declaiming with a mouth full of bread-pudding. "We can't win because we've no power. Might is always right, and we've not got the might. We're mice," he shouted, "threatening an elly-phunt, and the elly-phunt's going to jump on us with all four feet."

"Ouch!" Shanny said, and laughed.

Jek, his mind dizzy with whisky, had difficulty in understanding that Tom Shannon was talking about the strike. "We've got the Union," Tom Shannon went on, "but what's the Union compared to the money the Gaffers have got? And although the law allows us to strike, it has no sympathy with us, and no man can fight against the law. No, I'm afraid that we're doomed to failure from the beginning." He stopped and shook his head with admiration at his own words. "So a fellar was saying this morning in the 'Cross Keys' anyway," he said. "Very clever man," he told Jek, leaning forward and shaking a finger in his face. "Very clever man." Jek nodded a bleary agreement. Tom Shannon went on, "Meself, I don't think we'll win, because we never do. I never heard tell of anybody like us who beat the Gaffers. That's the way it goes. Can't win, that's just the way it goes."

Jek concentrated on what he had said, and came to the conclusion that it was against the strike. "Thee going to scab then?" he demanded fiercely.

The heads of all the Shannons, and all the Shannon dogs, came round to stare at him. "No," Tom Shannon said. "Thee've got to go along with the stream. There's a strike, we strike—an' wholehearted as anybody. But we won't win."

Jek flushed, and mumbled an apology to his jam-jar. Shanny got up, stumbling a little, and took a blanket from

25

the floor. "I'm going to bed," he said, and staggered towards the door. Jek got up to go with him, and found himself unsteady too. He went home, feeling cheated. The announcement of a strike, to his way of thinking, should have parades and trumpet-blowing, and angels on high. There had been none of these things and now it didn't seem real. He began to think that he'd imagined it all, or dreamt it last night. That was it. A dream, and tomorrow he had to get up at half-past three in the morning for fourteen hours of hard work.

He jumped into his own home and crashed against the settle. "Damn, I do that every bleedin' time!" he said.

The room was dark except for a red glow from the fire, and he thought everyone had gone to bed. But a voice from the fire said, "That thee, Jek?"

Jek peered towards the sound. "That thee, Grandad?"

"Ar," the old man said.

"I thought thee'd be in bed," Jek said. "Hang on, an' I'll help thee up the stairs." He moved forward, banged his shin hard on the edge of a chair, fell, and got the corner of the table in his ribs. He struggled up, swearing, climbed over the chair that had tripped him, and felt his way around the table to the hearth. "Come on, Grandad, up thee get."

"Wait, wait, wait," the old man said. "I've got summat to say to thee first."

"Oh." Jek crouched down at the old man's side, leaning against the range, and looked up for what he had to say.

The old man's face loomed over him like a gorse-bush, with its beard and bushy eye-brows. "This strike, Jek—tha knows it ain't no good?"

"Thee an' all?" Jek said. "Everybody's tellin' me it ain't no good, bar me Dad."

"It ain't no good," his grandfather repeated. "It'll have thee damned to Hell and the eternal flames and darkness."

Jek gasped and stared up at him. Dewi never showed any

26

signs of having heard that 'Christ' was anything but a swear-word, but Grandad Ellis was a Methodist, and he believed every word in the Bible; and he believed in Hell too. Jek had difficulty believing in God, but the idea of Hell, the eternal flames, the gnawing worm, the vile darkness, had him fascinated, and also terrified that it might be true. "How can there be fires and dark at the same time, Grandad?"

"By God's will," the old man replied. "Anything that God wills can be done—an' He won't be pleased over this strike, I can tell thee."

The pitch-black room, reddened by the fire, was like Hell. "Would He really send us to Hell for going on strike, Grandad?"

"All them as disobeys his laws are damned," said the old man. "An' this strike is against His laws. It's dead against His laws. God made all of us, Gaffers an' men. And He made the Gaffers Gaffers, and He made us like we are—now thee'm trying to alter that. Thee'm trying to rise above tha station, an' that's wrong, in God's eyes, that's a sin. Thee should be obeyin' tha rightful masters, that God chose and set over thee, to rule thee according to His will—an' here thee am, tryin' to fight 'em. When thee fight tha rightful masters, then thee'm fightin' God Hisself—and He'll damn thee for it."

Jek lowered his head to think about it, but he was confused. "It's true what I tell thee, Jek," Grandad Ellis said. "I'm older than thee, an' I know."

"God wants us to live like this, Grandad? He don't want us to have anything better?"

"I can't explain what He thinks, or why. But if He put thee to live like this, then this is how He wants thee. If He wants thee to have a pay rise, then He'll put it into the Gaffer's mind to give thee a pay rise, not before."

"I thought God was supposed to love we!"

27

"So He does, but He has His reasons for doin' everythin', an' He ain't going to like thee tryin' to fight Him. I'm tellin' thee, Jek, if thee value thy immortal soul, thee'll have nothin' to do with this strike."

"But that'd mean blackleggin'. I can't do that."

"Don't fear thy fellow-man, fear only the Lord God, your maker."

"It can't be wrong to want a bit better than thee've got," Jek said. "It can't be."

"It is. It's sinful and the Devil's work. Thee should count thy blessin's and say, 'Get thee behind me, Satan'."

Jek grew angry and shouted, "Well, I can't tell the difference between me thoughts that am sent by the Devil and them that am sent by God!"

"Hush up," said the old man. "Thee can give me a hand upstairs now."

# 2

He drifted in and out of sleep for a while, and then he was fully aware and blinking at the holes in the roof above him. It was very hot.

The three of them, Jek, Aynoch, Joe, lay on a straw-filled palliasse, thin and hard, laid on the floor. They slept naked and uncovered because of the heat, but crowded together because the palliasse was so narrow.

There was only one room upstairs, and all the family had to sleep there. So the boys' end of the room had been screened from the rest by a piece of old curtain, strung from wall to wall on a length of string. Jek looked up the height of the curtain as he lay, at its tired, frayed weave and faded blue. It wasn't washed very often, because in summer water was scarce, and in winter it was hard enough to dry clothes without adding to the job. The curtain smelt of must and dust and age. Its colour was depressing. Jek looked away from it.

No work, he thought suddenly. No work! It was a thought which did a lot to cure the Monday Morning Sickness. The thought of the strike was better still. Wouldn't they show the Gaffers! He pulled on his trousers, stepped into his clogs, took his shirt by the collar in one hand, and the chamber pot in the other. He stopped where the curtain met the wall, and called, "Mother? Sarah?" There was no answer, so he pushed through the curtain and made for the bedroom door, dutifully keeping his eyes away from the window side of the room where his mother and sister would be sleeping, probably naked. He would have liked to look—just from curiosity, never having seen a naked woman—but he didn't look because something in him wouldn't let him. It could have been duty, or a sense of right and wrong; but he thought it was more likely to be fear of the Devil, or some unknown disaster that would fall on him if he did.

Nellie came running to him, dancing up and down, chattering, undressed as yet. He opened the door with his shirt hand, and she clung to his legs all the way downstairs, making him splash the contents of the chamber pot about. Mice ran from the table and the range as they approached, scores of them, all scuttling across the tiles to their holes.

Jek put the chamber pot on the table, took his belt from the back of the door, and buckled it on. He remembered his grandfather's words about God, and Hell, and the strikers being damned. In daylight it wasn't half so believable, and he tucked the whole argument away in the back of his mind. The first thing he wanted to do was to wash the sweat-grease off himself, so he struggled back to the hearth and picked up the enamel bowl, to fetch some water from the Baden Brook.

Nellie came running across the track after him, still naked, and Jek laughed. Well, it didn't matter, she was only a little girl.

There weren't many people about; it was still early and blue, cool so that he shivered. Even the factories hadn't started yet. His footsteps echoed and his voice bounced back to him, lonely, from the Baden's high walls. The brook water was very cold, and chilled his fingers as he dipped the bowl. He splashed some suddenly over Nellie, making her jump and squeal. Then they went back, Jek swinging Nellie into the house first, then putting down the bowl and jumping in, crashing against the settle, patiently getting up, retrieving the bowl and placing it on the table.

30

He found that he'd forgotten to empty the chamber pot. He threw the contents of the pot outside, and the pot under the table. He left the door open, hanging inwards on its hinges, since it was going to be another fine day.

He lit the fire with coal stolen from the White 'Oss Pit, and wood from old boxes. Fire-lighting was a man's job, and he cursed this tradition as he wrestled with the ancient tinderbox.

Chore done, he could wash himself from head to belt with cold water and gritty soap that had been used to scrub the floor. When he dried himself he felt human, and not like something slimy. Then he seized Nellie, put her bodily into the bowl and gave her a bath, splashing water all over the table and floor, and getting soaked himself. Kneeling painfully on the uneven tiles he dried her well on a towel taken from the brass rod along the mantelpiece. He had seen his Grandad's hands split to the bone from chapping, and he had a horror that this might happen to soft little Nellie if he didn't make sure that she was perfectly dry. Nellie wriggled, blew raspberries and spat at him—and he laughed at her, ducked and made a game of it instead of slapping her face as he should have done. But the whole thing was a game. If Reenee had asked him to bath Nellie he would have indignantly refused, because that was woman's work, bathing children.

Breakfast. The breakfast things had been laid on the table the night before, by Reenee. A collier gets his own breakfast, because he believes that to meet a woman on his way to work is unlucky, like hearing a bird sing near the pit-head, or seeing a hare. If he does meet a woman, he must take a day off and lose a day's pay, or go to work and live in fear of an accident, either that day, or the next, or the next. . . .

There was half a loaf of bread, a carving knife, and a bowl of dripping. In winter there would be porridge, but now it was summer and water was scarce. Jek looked into the milk-jug. There was only enough for drinking.

"I want a story," Nellie said, slapping at his legs.

"Eh?" Jek asked, to infuriate her. He spread the top of the loaf with dripping, cut a thick slice, and folded it into a sandwich for her. The dripping had a dervish dance of mice-footprints in it, and the marks of their teeth. They had been nibbling at the bread too. But then, they nibbled everything, even some of the stuff hung from the ceiling. There was nothing to be done about it. He climbed up on the settle to fetch two cups down from the shelf, filled one with milk and put it on the corner of the table for her. Then he went to sit on the stairs, but Aynoch was stamping down them.

"What thee done with the pot?" Aynoch demanded.

Jek pointed to where it lay under the table. Aynoch studied it morosely. "Thee reckon the boggin'-hole'd be full if I run up to it?"

"There wasn't nobody by it as I could see when I went out," Jek said. "There's a bowl of water there to wash thysen when thee come back."

"Right," Aynoch said as he hauled himself up to the level of the track outside. "Mornin', Jim!" he shouted to someone out there. "Hey, Jek, here's Jim Woodall goin' to see the Gaffer."

Jek crashed and climbed around the chairs to get to the door and see. Jim Woodall was the White 'Oss's Union Man, their representative. He was going to inform the Gaffer that his employees were on strike from today onwards, until they were granted an increase in wages of one penny for every tub filled with coal.

"Good luck, Jim!" Jek shouted.

But Jim Woodall didn't hear.

\*    \*    \*

Long, long day. Without work there was nothing to do.

Reenee and Sarah were washing, using water from the Baden Brook, since that was all there was. They borrowed the maiding-tub and the scrubbing-tub from Mrs. Jones— the two families shared the tubs—and employed Aynoch and Joe in carrying water from the brook. It took an hour or two to fetch enough water to fill the maiding-tub, and to boil it up in saucepans and jars.

The tubs were made from a barrel. The top quarter of the barrel had been sawn off, and that was the scrubbing-tub. The remaining three-quarters of the barrel was the maiding-tub.

When the water was boiled, and the room filled with steam, it was carried outside to the maiding-tub and poured in. Soap was grated into the water, clothes were pushed down into it, and Reenee took up the Maiden. The Maiden was a heavy block of wood with a cross cut into its bottom, so that it had four legs. A long, broom-stick handle, with cross-piece, was attached to it. To clean the clothes, the Maiden was pounded up and down on them at great speed. Reenee could keep up the pounding for five minutes together but had no intention of doing so. Sarah had to do the maiding, and when Sarah got tired, Joe had to take over.

When the clothes came out of the maiding-tub, they were taken inside to Reenee. She had the scrubbing-board wedged between the bottom of the scrubbing-tub and her body. She spread the clothes over the board and scrubbed them vigorously with the scrubbing-brush until her arm felt as if it was dropping off; then she scrubbed some more.

Dewi went to the Tavern, where he drank slowly and watched through the window for Jim Woodall. Grandad Ellis's stool was put outside for him, so that he would be out of Reenee's way, and Aynoch and Jek had to remove themselves. Aynoch went off somewhere with some friends, but Jek didn't go with him because he didn't feel like company. He wanted to think. That was impossible near the houses, because of the women washing and gossiping, the old men, the dogs, the children. . . .

There was only one place in the Row where you could get any peace—and it wasn't the boggin'-hole, because with twenty-two, mostly large, families to use it, the boggin'-hole was in high demand. The peaceful place was the field. Jek walked up into the field until he was well over the slope and out of sight of the Row, past the quarry. He sank down below the grass-heads, to sit cross-legged and meditate.

It was this strike. There he had been, thinking that a strike was the be-all and end-all of everything. They would strike, and the Gaffers would have to give in; they would get higher wages and all the better ways of living that went with higher wages, and everybody would be happy.

But apparently it wasn't as simple as he had supposed.

The Shannons said the strike would fail because that was the way it went. Everything was fixed in the way that it was, and any effort to change it was wasted. You just had to drift along and grab what you could—like a pheasant from a lord's estate. Jek scowled. It was easy to fall into that easy way of thinking, and difficult to find a way round it—no it wasn't. Things did change. When Dewi had been young children went down the pit at five, not ten; and women and girls worked underground too, pulling the trucks. Now only boys—or officially only boys—worked down the pits, although women still did heavy work on the surface. But he couldn't think of another example. The Row and the Pit were just as they had been when Dewi was a boy, except for different faces. There had been no repairs made, except those done by the tenants; no more precautions taken at the Pit.

Jek listened to the trip-hammer pounding away at the

forge. That would be stopped soon. The White 'Oss Pit delivered coal daily to the forge, and to other, smaller factories. But not now they were on strike. The Gaffer would find himself short of cash, and perhaps the factory owners would press him to settle with the colliers and end the strike, so that they could all get back to business. Jek's hopes began to rise again, until he remembered his grandfather's talk of God's disapproval and revenge. Jek scowled. He didn't want to think of that.

"I don't believe in God," he told himself aloud. But that was rash—who knew? There was certainly, undoubtedly, a force for evil—so why not a God? And if there was a God, then He wouldn't take kindly to such remarks. "There might be a God," he added, to make amends.

He felt that his head was crammed to bursting point with ideas and conflicting opinions, and that the thin chain of thought that *was* Jek, was being pushed under and smothered by all the others. He got up hurriedly and ran back down to the Row, arriving in the stink of the muck-heaps hot and sweaty, so that all the flies came buzzing round him. He ran on, flailing his arms to keep them off, thinking that he would find Shanny and some others, and get up a game of Little Jack Appny. By the time that was finished, he would be too tired and hot to think of anything else for the rest of the day.

Little Jack Appny is played like this: one player is elected the Block of Wood, another the Chisel, and two or three more are the Hammers. The Block of Wood stands against a wall, and the Chisel bends down, as in leap-frog, with his head in the Block's stomach. The Hammers adopt the same position behind the Chisel. The other players—as many people, male or female, as can be found—run at the line of Hammers, Chisel and Block, and jump on to the Hammers' backs, shouting, "Little Jack Appny, up the stick, over the gate, two, four, six, eight, and off again

35

Bodger." Those who jumped short hitch up along the Hammers' backs, to make room for those behind; and those behind climb over those in front. The point of the game was supposed to be to see how many people could be piled on the backs of the Hammers before they all fell down; but it was played more for the pleasure of running and jumping, seeing the Block's face as the Chisel's head was knocked into his stomach, and all falling down in a heap.

He found Shanny easily enough, and they found plenty of others to play the game, but they could not find Dai Macnamara, nor Ayli Black, whom they particularly wanted.

The bruising, screaming game of Little Jack Appny ended abruptly when Jim Woodall came back, in the early afternoon. They knew that he'd come because of the sudden movement of the men sitting on the field bank, the sudden chatter. Snatching up their shirts, they ran to the gathering crowd at the other end of the Row. But the crowd broke and scattered before they got there.

They stopped men, catching at arm after arm, asking, "What did he say? What did he say?"

They were told, "Gaffer wouldn't see him. Heard what he'd come about and said he'd see him on no account. Jim ses he tried to argue, but they wouldn't have it, wouldn't let him in. Reckons he tried for nigh on two hours, until they said they'd fetch a copper—then he come home."

That was dispiriting. The game of Little Jack Appny didn't continue, and Jek wandered away from the others. He went to sit on the low field wall by the road. He dangled his feet, kicked his clogs against the brickwork, gloomily watched the carts go by, and was miserable. He was very easily made miserable, he thought, and wondered whether the feeling was genuine, or only play-acting. And play-acting made him think of the plays put on at the Palace of Varieties; and thinking of the plays made him remember

36

the way the heroine always cast herself into the hero's arms at the end, to the cheers and whistles of the audience; and that made him think—naturally—of Ada Bishop.

He thought about going to see her. He rubbed his ear and seriously considered going to see her. She would be in, because she didn't work, she stayed at home to help her mother. Her father and brothers were in the iron trade, which was doing well at the moment, and they earned good wages.

A good-looking wench was Ada, and it was noticeable that out of all the lads, it was Jechonias Davies she most chose to talk to, and walk along the Oldbury Road with. That was reason for pride.

But by degrees he gave up the idea of going to see her. Good-looking she was, but irritating. Sometimes he disliked her. She was just like Shanny, she didn't listen to what was being said; her mind ran on its own daft girl's matters, and when you looked to her for an answer or opinion, she said something about sewing, or onions.

Then his wall-kicking and wondering was broken by the arrival of a large carriage at the end of the Row, amongst the muck-heaps, and the flies, and the crumbling houses. A gentleman's carriage. The Gaffer's carriage.

Jek stared at it. He had seen the carriage before, twice before, at the Pit, but seeing it here was like having God himself walk in and ask: Could he borrow a cup of oatmeal until Friday. The men came from the Tavern steps and from the fields. Jek jack-knifed from his seat on the wall, and hurried over to join them.

The Manager of the White 'Oss Pit, neat and well-dressed, stepped down from his carriage and looked around him. He saw no reason to hide the disgust that he felt. Couldn't these people—he always thought of them as 'these people'—keep the places they lived in clean and in good repair? Less like a pig-sty?

37

The houses were filthy and rickety. Rubbish and—horrible—filth lay in stinking heaps on the dirt-track in front of the buildings. He found it almost impossible to breathe, the stench was so paralysing. It made his stomach heave. All this proved the point that it was of no use providing these people with decent homes, since they would only reduce them to slums in a matter of months. There—in the gutter, quite close to him—a dead dog, crawling and buzzing with big, black, shiny flies. And there *they* stood, in front of him. They breathed easily enough. They must be quite immune by now; after all, they had left the rubbish there in the first place. They must like it. They must find the air quite bracing.

He regretted now that he had come to speak to them in person—he could have sent one of his foremen . . . but no. It was best that he should speak. He could put the case so much better than any ignorant foreman, and his words would have more authority.

He looked at them, wondering how to begin. They stood in front of him, in a rough half-circle, heads low, watching him. They were surly and dirty, like all of their class. They smelt. And their clothes were so odd. Jackets too big and jackets too small; all jackets too short, reaching only to the hips. A lot of the men were in their shirt-sleeves, some of them even naked to the waist. They had no sense of decency at all. Ties they wore, but no collars. Flat caps, of course, worn straight, to the side of the head and with the peak at the back. A few wore heavy pit-boots, quite a few of the younger ones went bare-foot; but those who really made him smile wore clogs. He had thought only the Dutch wore great clumping clogs.

Here came the women. The swearing scolds in men's boots and caps, with shawls over their heads and aprons around their waists. They were worse than the men. So much more—common.

38

He opened his mouth to speak, and drew in a lungful of the thick atmosphere. He began to cough and choke. The colliers clapped slowly, with surly, watching faces.

His mouth tightened. He'd stop their clapping. "A Mister James Woody called on me this morning, to inform me that the colliers of the White Horse, among others, have decided to stop work until the owners and myself grant you a pay increase of a penny a tub."

The colliers were silent. He was distracted by the crash, and clatter, screech, bang, howl and buzz of factories all around him, and it was several seconds before he went on. "You all signed—or rather, put your marks to—the April Agreement. Perhaps you didn't understand what you were agreeing to—I'll explain. The April Agreement is a contract between you and the owners. You agreed to work for the owners until next April—*on their terms*. They agreed to provide you with housing and tools." He was pleased to see some nervous faces amongst the crowd, some worried glances, one to another. Any mention of the law always frightened the common people.

"You will understand," he continued, "that the pay increase you 'demand' is impossible. Why, it would mean, for most men, a weekly increase of ten shillings. . . ."

"I'll buy up the Pit with that," a voice shouted.

The Manager lost his fatherly tone, and frowned. "An increase of ten shillings. If you were to succeed in *forcing*—blackmailing, I say—the owners and myself to pay this huge sum to all of you, in addition to your present wage, you would very soon have no job. The Pit would simply have to close down. You will understand this, my friends. And if, *if*, it were at all possible to find the resources with which to pay you—then, no doubt, the women who work on the Pit banks would want an increase in their wages . . . and so it would go on. You will understand this, my friends. *Now*, I want an immediate return. . . ."

39

A man stepped forward from the other colliers. A short, slight man, whose trousers concertina-ed down his legs, whose jacket was baggy and fraying at the cuffs. He respectfully took off his cap, to show a thin, stubbled face, still young-looking despite the deep lines at the eyes and brow. He seemed far too small and worried to hold a position of authority, even in a colliers' 'Union'. He said, "If I could just say-say-say a word or two, please sir," with a stammer and a hesitating over the words.

There came murmurs of approval and encouragement from the colliers behind him. "Thee tell him, Jim, go on, Jim, tell him, go on."

The Manager frowned at the man and asked coldly, "Mister Woody?"

"Wood-Woodall, sir, not Woody. Sir—we—we understand your p-position, sir, but, please sir, it's not greed. We *need* the money, sir. Prices are going up all the time, and when the women have bought the groceries, sir, there's nothin' left for clothes, sir. Sir—the childer run about without no shoes as it is, but they can't go without shoes, sir, especially with winter comin'; an' funeral clubs, sir . . . we need the funeral clubs—an' twopences for schoolin'—an'—an'—"

The Manager lost patience. "Forgive me, Mister Woody. I am not interested in your domestic arrangements." What were they doing spending pennies on 'schooling', he wondered.

"Yes, sir, I'm sorry sir," James Woodall went on, "but I was trying to tell you. With—with *respect*, sir, we don't think, we don't think, sir, that a ten-shillin' advance is goin' to break the owners. Sir."

The Manager gasped at this audacity, this flat denial of the words he had spoken only a minute before. The colliers, who had been listening dubiously to Jim's respectful speech, gave a shout of pleasure.

"Mister Woody," the Manager said deliberately, "I hardly feel that you, a face worker, can be qualified to tell *me* about the Pit's finances. I assure you that to pay every man here another ten shillings a week, over and above his present wage, would ruin the owners—and I believe that you are proposing that every man who works for the White Horse Pit should receive that amount."

Jim Woodall replied, stammering, but firmly, "Sir— we-we-we think that w-what you just said is a lie."

The shock of this defiance, and the stink of the district, came near to choking the Manager. His voice died in his throat, and his face became red.

Jim Woodall pressed home his advantage. He hurried on, stumbling over the words, trying to hint that the colliers would settle for less than the amount demanded, without showing weakness. "We-we *need* more money, sir, car-car-can't take less—can't take less than a penny a tub more. Sir, we've asked before, we asked and asked, but you said no, so we've got to strike, sir. Sir-sir, we *need* the money, sir, we do."

The Manager found his voice and answered loudly, "Perhaps if you spent less of your money in the Public Houses and on 'schooling' you would manage better. Your wage is quite—*quite*—sufficient for your needs if you spent it sensibly, instead of wasting it."

There was a silence, reaching out from the staring colliers to the Manager. The factories still clamoured, but the noise from them seemed to have faded, to have become insignificant.

"Sir," Woodall said, licking his lips, "we can't live on the money we get. There's childer have never tasted meat. They live on potatoes an' porridge or groats. We try to give 'em some schoolin', sir . . ."

It seemed that this shabby little man was not so easily brushed off as you would think to look at him. The Manager

41

said, "You are repeating yourself, Mister Woody. And I have no desire to argue with you. I came here only to inform you that your strike is futile, and to tell you that I want an immediate return to work. If you return to work today, this will be forgotten. If you do not—well then, I shall have to take action. You rent your houses from the Pit on the condition that you work at the Pit. If you continue this strike, I can and shall evict you. I shall expect to see every man here at the Pit within an hour, and I shall expect you to make up for lost time. Do I make myself clear?"

There was the same silence. Then someone shouted, "Get stuffed!" and produced a little laughter. "Very well," the Manager said firmly, and turned to go back to his carriage. But James Woodall, that dirty, shrunken, stubble-faced little man, stepped up, and actually took hold of his sleeve. He drew back, but the man's grip tightened. The grey, worried face peered up at him. Woodall said, "Sir, you eat well every day, and you drink an' all; an' your Missis and your childer—they're well-fed an' well-clothed. Sir—you're a ge-ge-generous man. You don't grudge us the same, surely sir? An' we can't manage on what we get."

Suddenly furious, the Manager shook him off. "Thank you," he said sarcastically, as Jim Woodall stepped back. "Thank you for allowing me to leave. You have one hour in which to return to work."

"But sir . . ." that insufferable little man whined again.

"You Woody!" the Manager shouted. "You are sacked! As from today your services are *not* required." He restrained his undignified temper, drew himself up, smoothed the front of his coat, and patted his breast pocket where the comfortable round of his gold watch lay. "You will be out of the house you now occupy by Wednesday. If you are not, you will be evicted."

Jim Woodall stopped as though his clockwork had run

down, with little shivers and shakes, staring at the Manager from a sick face. The colliers stared too, all lifted up their heads and all stared at him. The Manager smiled. Thought he was bluffing, did they? He'd shown them!

Then, with a shock that remained with him, unfaded, for the rest of his life, a stone bounced off his forehead.

He suddenly realized that he was in a rough district, alone. It was unlikely that any divine intervention would save him if they took it into their heads to kill him. He whirled and climbed hastily into his carriage, and drove away, his coachman as eager to leave as he was himself.

Everyone looked at Jim. Seeing their eyes on him, he swallowed and said, with forced light-heartedness, "Well, we expected that, didn't we? It's all bluff, he won't do anythin'. No use givin' in now, is it? We gotta try, else we've no chance of winning."

"He's sacked you an' throwed you out your house," one of the men pointed out.

"Ar," agreed another. "What you going to do now? This is a right mess, ain't?"

They were blaming him, Jek saw. They were blaming Jim for what the Gaffer threatened to do. "This is a right mess, ain't?"—meaning, You got us into it, now get us out. Jek was disgusted, and was relieved to see his own disgust in other faces as they watched the two men who had spoken.

"Shut your gob," Dewi said, standing beside Jim. "You knowed what you was riskin'. You voted to go on strike. You looked good then. Don't show your real colours now and spoil it."

Seth Jones came to join Dewi and Jim. He said quietly, "Jim, if I was thee, I'd go an' tell Annie."

Jim nodded, and walked off up the Row in obedience to this gentle dismissal. Annie, his wife, was waiting for him not far away, hard-faced. "We gotta get out the house,"

43

he told her from a distance. She nodded, and smiled at him.

*   *   *

Some men had never joined the Union. They thought it too revolutionary, too dangerous. Or they had been taught by the church that man was called to his estate by God, and that it was a sin to try to alter it in any way. In Heaven there would be equality, riches and happiness, if you were good enough to reach it.

Others had joined because they had decided that it was a good investment. You paid your sixpence every week into the Union Fund, and then, if you were sick and not earning, the Union would give you enough money to live on until you could work again. The Union helped widows and orphans, and would give assistance in paying doctor's bills. If a man was really hard-up, and couldn't afford winter clothes for his family, he could go to the Union Man, whom he knew personally, and tell him so; and the Union Man would organize a whip-round, or give him money from the Union Fund, if the general opinion of the Union members was in favour.

And if there were any complaints, if a scales-man was suspected of weighing the coal against the colliers, if the stores were slow in providing a new pit-prop to replace one that was 'singing' or creaking; or in replacing worn-out tools; if no one listened to the colliers when they said that a particular gallery had become dangerous and should be closed off—then the Union Man tried to put things right. He went to the offices and argued with the people there; or he found the man in charge of the stores and tried to persuade him, peaceably, to speed matters up.

The Union Man and the Union were always there, to take your problems to at any time. It was very comforting to have such an arrangement for when you were really in trouble, and with all these demands on Union funds the possibility of a strike was far away.

But now the Union was calling a strike, and the condition of joining had been: In the event of industrial action, one out, all out.

Now they all had to decide. Whether to go back to work as the Gaffer ordered them to do, and lose the support of the Union and the friendship of those men loyal to the Union, because they were afraid of losing jobs and houses.

Or to stay with the strike because they couldn't face the contempt of their friends.

Or, like Jek, to be entirely on the side of the Union. He felt no doubt—only fear of what would happen if the Gaffer won. If. Jek hoped, wished, that the colliers might come through this fight without too much—trouble. Without his father or any more people being sacked or evicted. He wished for this until it hurt under his ribs, until he felt empty—that was his idea of praying, though he didn't know whom or what the prayer was directed to.

"Thee scared, Shan?" he asked in a whisper, because he was asking a personal question.

"Scared?" Shanny said, lying on his back at the top of the field bank. "Scared o' what?"

"Of what'll happen if the Gaffer gets wicked, what do thee think?" Jek snapped.

Shanny gave some thought to it, whistling. "No," he said.

Jek glanced at him sideways. "I don't believe thee."

"Don't then," Shanny said. "But I ain't much scared. See, what'll come will come, see? If somethin' bad's coming, then it'll come, whatever we say or do. If it wasn't the Gaffer having everybody sacked, it'd be the fever, that or the Pit'd fall in on we. No point in worryin' over it." He added, "We can look after ourselves anyroad. We'd

45

manage if—we'd manage if the sky fell down, we'd manage."

*   *   *

They knocked on the door and pushed it open. Dewi said, "Can we have a word, Our Kid?"

Jim Woodall looked up from the table where he was sitting, and nodded. So Dewi, Seth Jones, Methody Bates and two or three others crowded into the neat little room. Dewi gave a stiff nod of the head to Anne Woodall by the fire. "Won't be a minute, Annie." He coughed, and said to Jim, "We just come to say that we're sorry thee was fired an' throwed out."

Jim smiled. "Not half so sorry as I am."

"Well, we was thinking," Dewi went on, "we was talkin', an' it occurred to we as thee'd be lookin' for a new job?"

Jim nodded agreement.

"Ah, well then, don't bother!" Dewi said triumphantly. "Thee'm still in Union, ain't? Well, the Union's employin' thee as Union Man. We'll pay thee an' we'll make it a condition that unless thee get tha job back, we don't go back to work. How's that?"

"That's saftness!" Jim said. "That'll finish it. He'll never give in to thee if he's gotta give me my job back as well."

"He's not the owner," Seth Jones said, "he's only the Manager. He's got to make the Pit pay. Tha job won't make any difference, one way or the other."

"We'll put thee up when thee'm evicted," Dewi said. "An' we'll help thee shift the stuff out if tha like."

"No," said Jim. "I'm not raisin' a finger to move anything out of here."

Dewi grinned. "If tha like," he said, "we'll stop 'em from coming in, we won't let 'em evict thee."

"No," Jim said. He looked up again, grinned weakly and

46

said, "Look, I-I-I I'm Union Man, right? Well, I say, no fightin', no-no fightin' with anybody—pickets or Gaffer's men or a-anybody. 'Cos, listen, listen—it'll give the Union a bad name an' that—see, that'll be the end o' Unions." He chewed his lip for a while, and then said, "Look. I'll have me Union pay until I can get another job, 'cos I got no choice, an' thanks for that. But that business about askin' for me job back—no. An' I ain't being Union Man no more, 'cos with looking for another job I couldn't do it right. So thee'll have to elect a new Union Man."

They said, "No," unanimously. "We elected thee," Dewi explained, "an' thee'm still alive an' in tha right mind, so thee'm still Union Man. Thee'm still in Union an' when we win strike thee'll have tha job back."

"Thee talked beautiful to the Gaffer," Seth said. "We got nobody else as could talk like that—'swhy we chose thee in first place. Thee'm educated."

Dewi nodded solemnly and repeated, "Educated."

Jim was flattered. And he didn't want to seem to be backing out of the strike. He nodded and mumbled, red-faced.

Someone at the back of the committee said, "Look here. Where they goin'? They got their snap with 'em! Where do—"

Dewi peered out of the window at the hurrying men. "They'm black-leggin'," he said.

"Black-leggin'!"

"Pickets," Jim Woodall said. "We've got to get pickets out—but no fightin'. Keep it peaceful like."

They left the house and scattered, to search up and down the Row for sons, brothers and friends. Already some were missing. They had made up their minds to return to work.

When Jim went out he was approached by three colliers.

"Ado, Mister Woodall," said one, greeting him.

47

Jim blinked nervously. Since when had he been anything but 'Jim'? "Ado, Fred," he said. "What's up?"

"Nothin's up," Fred said. "We just come to tell thee that me an' Eben an' Simon here—we'm goin' back to work. We thought it was only polite, like, to come an' tell thee. So we know where we stand."

"Oh," Jim said. "Oh. Well—thanks, Fred, Eben. Simon."

"We heard stories," Fred said, "about other strikes an' what happened to them as went to work there. About how they was laid about with pick-'andles and their windows smashed an' their wives an' childer tormented. That's what we heard—so we thought we'd let thee know where we stand."

"No need to worry," Jim said. "No violence. Not in this strike. Peaceful picketing."

"Ar," said Fred. "We *want* to go to work."

"Ar," said Eben. "We want to."

"None of that here," Jim said. "I swear. No violence."

Fred glared at him steadily for a long time, then turned and walked away with his two friends. Simon, who had said nothing, kept looking at Jim over his shoulder and then came running back. He stood silently in front of Jim.

"I'm sorry," he said, at last.

"No nee-nee-need to say sorry, Si," Jim answered, flustered.

"I'm sorry. But I can't—I gotta go back, see. There's Our Mariah, an' the childer, they gotta have house, an' what if I got the sack? The eldest's only five. Tha see. . . ."

"Si!" bellowed Fred.

"I'm sorry," Simon said, and ran after Fred and Eben.

# 3

---

The Manager dropped some fish on to the china plate he held, added an omelette and some breast of chicken, and passed the plate to his wife, who was seated at the table. For his own breakfast he chose bacon, eggs, a pair of kidneys and—after some humming—broiled chicken. He took his seat opposite his wife. He preferred private meals like this to be informal.

"Most of them *already* earn ten shillings a week, and yet they have the audacity to ask for *twopence* a tub."

His wife smiled pleasantly. She was very pleasant.

"Twopence a tub would make their weekly wage a pound. Can you imagine! They would spend it all in drink, and the streets wouldn't be safe. We have already seen what happened in France—the King beheaded, the Queen too, the rabble ruling. . . ."

"Oh, such a long time ago," his wife said, in slight protest.

"That is the purpose of History, my dear—to warn. We have our warning. Twopence a tub!"

Around them, reflected in the dark, polished wood of the table, were little cut-glass dishes, glinting in the sunshine from the window: filled with fresh strawberries and pre-serves; and cream. Delicate glass plates, catching the light and holding it, were arranged with fingers of bread and butter; soft bread rolls. A stand of boiled eggs; a muffin dish full of buttered toast.

"Never mind," the Manager went on, more calmly, wiping his mouth before drinking. "I'll soon have them back at work. I sacked their leader. That will make them think twice before causing any more trouble. They'll soon be very glad to return to work."

The dark, polished sideboard was covered with a harshly white cloth. On it were laid the silver-plated knives, forks, spoons; the translucent china and the china plates, painted with fine sprays of flowers and rose petals. And the tongue, the ham, and the cold veal pies.

"Really," said the Manager, "they have a wage perfectly sufficient to feed and clothe a small family, and yet they think nothing, nothing, of eleven children. Surely it must be obvious to the stupidest of them that you do not have more children than you can afford to keep." He spoke with a little spit of jealousy. He had no children and he could afford to keep them . . . three pretty daughters and a son. "But there—they have eleven children, find they cannot keep them and demand more wages. They should count their blessings and be thankful for what they have." He poured himself another cup of coffee, added cream and sugar. "My dear, do you think that the cook could produce one of those iced, cream gateaux for tonight? I feel that I deserve a treat."

The polished side-table held the hot food. Poached eggs, grilled bacon, dressed fish, kidneys, lamb cutlets, tender broiled chicken and a dish of ham-filled omelettes.

And when they had finished, the servants came in to clear the table and wash the dishes.

\*     \*     \*

A crowd stood in a semi-circle outside Jim Woodall's house in White 'Oss Row. In the mouth of the semi-circle stood Jim Woodall, and his wife, Anne. Behind them was nearly every other person who lived in the Row. They all watched silently as the Gaffer's bailiffs emptied the Woodall house.

Jim, as he had promised, was giving no help in the eviction. He stood, hands in pockets, and studied the curved toe of his clog. But Anne had gathered up all her cups and saucers, which she had taken great care to keep matching; her pots and pans and stewing-jars; her dishes and plates, her jugs and knives and apostle spoons; and her fairings from the mantelpiece. The fairings were better than most, and she held them in her arms, with the crockery piled about her feet. She hadn't anything so hideous as the Davieses' Highlander. She hugged two black and white china dogs; a glass swan; a little flower-covered vase; and a china group of a man and woman climbing into a curtained bed, with the inscription, "The last one into bed turns out the light."

The bailiffs first of all brought out the chairs, and they brought them out carefully, setting them gently on their legs; because they knew why the colliers had turned out to watch. The table had to be dismantled in order to bring it through the door, but two of them carefully laid down the top, and the third put the legs beside it in a neat bundle, looking nervously up at the colliers as he did so and showing the whites of his eyes.

The oldest of the bailiffs came to the door with the real woollen rug that Annie had saved patiently for. He threw it on to the table-top, to save coming out, and gave the colliers a quick glance, to dare them to say that this was unreasonable.

The colliers were touchy, and there was some shuffling and looking at each other. But Jim Woodall straightened up and cleared his throat. The colliers slowly subsided. No violence had been decided on, and no man wanted to be the first to break that ruling. Jim watched his toe again.

The oldest bailiff came out again with the curtains from

the upstairs and downstairs windows, the picture of the lady from the kitchen wall, and the 'Bless This House' prayer. He put them on the rug and went back inside.

The palliasses were brought out next, with the sheets and the clean sacking blankets. The bailiffs laid them on the ground, and then the house was empty. Two of the bailiffs walked slowly away down the Row, ashamed to hurry, and afraid to stay.

But the oldest man stayed to shut the door of the empty house and to drop the latch. Turning, he nodded to Jim, and said curtly, "Afternoon."

Jim nodded back, said, "Afternoon."

The bailiff strolled casually away.

Jek thought it was pathetic that all Jim's and Annie's home should be thrown out on the track like that, in a heap. Annie always had everything so neat. All her cups and saucers were yellow; she had print curtains for the windows and pictures for the walls; only the best ornaments, sheets for the beds and a proper rug. His mother said Anne Woodall was 'high and mighty', but Jek thought she was womanly.

Anne Woodall, turning with her arms full of fairings, saw him staring at her, scowling and concerned, and she had to smile at him. He continued to scowl, with confused thoughts of play heroines and worthy women whom Jesus raised from their knees.

Dewi said, "Isaac Walters'll let thee keep tha furniture in his stables, sure to. Thee want to get 'em inside, 'case it rains."

After a pause, Jim said vaguely, "I'll go an' ask him."

"No need to ask!" Dewi said. "Thee know old Isaac'll let thee. He'll reckon it's good for business. Come on."

The colliers humped palliasses, the table-top and chairs; and began a procession towards the Tavern pub, Isaac Walters and the Tavern stables.

Jek saw Anne Woodall struggling to pick up her crockery and went quickly to her.

"I'll carry 'em for thee—if tha like," he said, and nearly choked with shyness for all he'd known her all his life. He gathered up pots and pans and crockery before she could answer, fitting one thing inside another, carrying more than was practical to show that he could. He followed her down the Row to the Joneses', where she was to lodge, giving her such pitying, sympathetic looks, that she laughed in spite of everything.

# 4

There were strikers and strike-breakers—blacklegs. The blacklegs went into the Pit under the protection of the 'guards' hired by the Gaffer. The 'guards' had been recruited around the pubs and lodging houses of Dudley; but the Gaffer was perfectly within his rights. He was protecting the property of his employer.

Jim Woodall's pleas for no violence became more urgent. He feared that the colliers might get badly hurt if they started a row with the guards. But there were scuffles between the two sides, it was only to be expected. Sometimes the colliers were doing no more than 'peaceful picketing', sometimes they deliberately picked a fight.

The scuffles became more and more vicious. To begin with, the colliers had armed themselves with pokers, and chair-legs, because the guards were armed. The guards then looked for more offensive weapons, because the colliers were armed. The colliers then looked for, or made, still more dangerous weapons—and the Gaffer took the first opportunity to call in the police to protect his employer's property. When the police arrived, the colliers scattered and ran, but some of them were arrested and fined for breaking the peace. The Union had to pay the fines, and this reduced the Union's funds.

So the picketing ended, and because of Jim's constant appeals for peace, there was very little other discouragement of the blacklegs. The colliers sat about, or found odd jobs; and the Shannons poached.

But what was the use of striking if the blacklegs went in and did the work? What quicker way was there of losing the strike?

Dewi went to Jim and told him that the blacklegs had to be stopped, one way or another. He said plainly that the majority opinion was for gutting and crippling the blacklegs, and that Jim was all that stood in their way. He told Jim that the rest of the colliers, including himself, weren't going to listen to one man much longer, and that when they stopped listening, Jim Woodall was as likely to get hurt as any blackleg.

Jim answered with his usual argument about keeping the Union's good name, all Unions, in the future as well as now; and Dewi very reluctantly agreed with him. He went home, and said nothing to anybody except Seth Jones, and that in his own house. He made sure that his family wouldn't repeat what he said.

No one wanted to be the first man to break the no violence rule.

Then came Timothy Russel, with his five brothers, Shadrach, Caleb, Tommy, Harry and Eric. They arrived at the White 'Oss Row with their spare shirts and bedding wrapped up in bundles, having walked eight miles from the Jubilee Pit. They brought news that the Jubilee, the Red Lion, the Hailstone and the Barleycorn had all joined the strike. But their family at the Jubilee Pit were strike-breaking, so Timothy Russel had led his brothers to the White 'Oss, because they had relatives there.

But the White 'Oss Russels were strike-breaking too. There was a big family argument, which everyone in the Row enjoyed, and then the brothers stood in the middle of the track, cursing their family, and God, and the Gaffers. But plenty of people were willing to take them in.

Jek was beginning to feel very unsettled. He had a troubled feeling in his stomach, and a confusion in his head that wouldn't let him sit still. It was all right when he had a job to do, when he was helping the bargees leg their narrow-boats through the canal-tunnels for a few coppers, or hunting for fire-wood to sell, but when he could find no jobs, and was idle, then he was tormented. He roamed aimlessly about the Row, the streets, the canal, the field, and grew bad-tempered; kicking dogs and throwing stones at cats. He got no real pleasure from making the cats or the dogs squeal, but he had to hurt something, and it was easier to hurt an animal than a factory wall.

He was on the side of the strikers. No doubt about that. Whatever happened, he would be on the strikers' side. Because he had said he was and he could never go back on what he'd said. But he was no longer sure whether he was right.

They'd come out on strike, and yet most people seemed to think that they'd lost before they'd started. Money was short, food was shorter, and that, as Reenee said, made the women and children suffer. And a good many men among the strikers didn't want to be on strike, they were just afraid to be blacklegs.

Then there were all the people who had to stop work because the colliers stopped work, although they weren't even members of the Union. The Pit-bank wenches, who carried the dirt and stone from the pit to the tips, and the masculine women who rode the horses carrying the coal to the canal barges; the sorter and washers who worked in the sheds above ground; and the factory workers and foundry workers. Some of them were still working, because of the blacklegs, and because not all the pits were on strike—but most of them were laid off.

It had occurred to Jek, although he didn't like the thought, that the strike-breakers were only following their own minds, just as the strikers were. Thoughts like this made life difficult. It was much easier if you could believe absolutely that strike-breakers were a mixture of the Gaffer and the Devil. Then it didn't matter what you did to them. But suppose they were just men, who believed in what they were doing, just as you believed in your views?

Timothy Russel had suggested sawing and chafing the rope that went down the North shaft into the Pit; so that when it was next used, it would break, with luck, providing the fraying wasn't noticed.

Jek could still scarcely believe that *that* had been put forward as a serious suggestion, and—even worse—that a good many of the younger men, including Shanny, had agreed with it.

It was murder, and worse. This district was known for accidents in all industries—arms lopped off by machines, bodies crushed under toppling heaps of castings, or sliced through by white-hot snakes of steel. And the pit accidents: Cave-in: seventy-five men killed in a few minutes of time. Explosion: eighty-two men killed. Flood: forty men drowned and crushed, twenty-seven bodies never recovered. Wildfire: sixty-five men and boys roasted alive. Gas: innumerable men suffocated as they worked. Men crushed by trucks, killed by shaft-ropes breaking, trampled and maimed by pit-ponies in the narrow galleries. The numbers were countless.

But when there was a cave-in, or a flood, or an explosion —then work stopped; not only in the Pit where the accident happened, but in all Pits around; men walked ten miles in an afternoon, to see if they could be of help.

To think of causing an accident was—unbelievable. The strike-breakers were only following their own minds. And if that made him a 'little snot-nose', as Tim Russel had said, well then, he was a little snot-nose and who the hell cared?

Jek did.

There was no middle road. You were a striker or a strike-breaker. You couldn't say, 'I'm for the strike, but I don't reckon this about beating the blacklegs up', because then the strikers accused you of not being on their side. As when he'd tried to explain what he thought about the idea of stretching throat-high ropes across the Pit entrances, in the hope that the blacklegs would break their necks. A great howl had risen about him. He'd tried to shout them down, or talk above their hooting, but it was no good. There was no middle road.

He had slunk away, and he thought of Ada again. She'd give him sympathy, if not understanding. He could shout his anger at her, and she'd say, "Oh, shame," and get angry at the people he was complaining about. Taking his problem to her wouldn't serve any useful purpose, but it would be satisfying.

He hadn't been to see Ada for a long time, though. She'd be niggly over that, because she didn't like to think that anybody could prefer anybody's company to hers. So he wandered over the fields, and found some late coltsfoots in the shelter of the brick wall around the pig-field. He picked them all, and put them in his pocket, so that no one would see him with them. A movement caught his eye, and he saw a cabbage-white butterfly perched on the flowers of a meadow-sweet. They didn't see many butterflies, not even cabbage-whites. He grabbed for it, but it flew away, to land again in the same place. This time he caught it, with great cunning, in his cap. Armed with these presents, he went to see Ada. He'd ask her if she could come for a walk along the canal.

He stepped up to the Bishops' doorway and rapped on the door. He whistled as he waited, and cupped his hand about the flowers in his pocket, ready to take them out when Ada appeared.

58

The door was opened by—Mr. Bishop. Jek was surprised for a second, then he remembered that the foundry would be on short weeks because of the coal shortage from the strike. But men rarely answered the door.

"Mr. Bishop! Er—is Ada in? Can I speak to her please?"

Mr. Bishop turned back into the house, bellowing, "Ada!" Ada left the fire, where she was peeling potatoes, and came to the door, filling it. She set her hands on her wide hips and stared down her nose at Jek. She was dark, and 'handsome', and she made Jek look insignificant. But he had too much contrariness to sink beneath her shadow.

He chose not to see her bad temper and held out the bunch of coltsfoots. "I brung thee these."

Ada looked poisonously at the flowers and said nothing.

Jek's anger began to spread to her. He didn't like being treated to Ada's little moods, especially at the moment. All right, so he hadn't been to see her for a week or so—hadn't she gone out for a walk with Tommy Allsop? Was he giving her the silence and the pout for that? He was not. He said, "What's up, Ada? Let's have none of the long faces."

She said, "A fine one, you are."

"What's that supposed to mean?" Jek demanded.

She leaned forward and hissed, "You stop away for months on end, and then you come crawling round here. . . ."

Jek swung his arm and threw away the flowers. "I ain't beggin' for owt off *you*."

"What you come for then?" Ada asked sneeringly.

Jek could find no answer to that, since he had come, cap in hand and bearing gifts, to ask her to walk out with him.

Ada suddenly whisked around. "I got work to do, even if *you* haven't!" she said. Jek had just time to see the grinning face of one of her brothers before she slammed the door in his face.

It was a conspiracy, that's what it was. All the Bishop

59

family were against him. Because the foundry was on short weeks through the colliers' strike and because he was a collier. Now he had no one to talk to. No one who would even pretend to be listening.

He remembered the butterfly, and unfolded his cap to free it. But he had damaged its wings and it couldn't fly. It fell to the dried earth road and struggled there. He stamped on it, reduced it to a damp patch in the dirt.

He crossed the street and wove in and out of the muck-heaps, shaking his head at the flies.

He wanted to talk to somebody, somebody who would *listen*; and tell them what he thought about the strikers and the 'breakers and everything, and ask them, did they agree?

Dewi wouldn't listen, because Dewi thought he was a child, and no child had anything to say that was worth listening to. You had to be forty years old before Dewi would lend even half an ear to anything you said. And Reenee never listened to anything except her own nagging voice. Joe and Sarah and Aynoch would only laugh. Shanny would wait until you had finished talking, then say something which proved that he hadn't been listening to a word you'd said.

There was Aunt Ruby. She'd listen, but only in order to make soothing noises at the right time. He wanted soothing noises, but he wanted somebody to take notice of what he said too.

It was maddening. He kicked out hard at a muck-heap; his foot sank into it and flies rose around him, buzzing.

He hissed between his teeth and kicked up brown water from the Baden Brook. He wanted to destroy something, to smash something, and the feeling frightened him.

He turned and ran around the end of the Row, under the overhanging house. He would go and chase the piglets and kick the sow. Especially the sow. She'd make him hop and

jump. With her after him he'd soon forget daft maunder-
ings.

Grandad Ellis was moodily watching Aynoch and his
friends playing leapfrog. "Energy to waste," the old man
muttered. "Runnin' around. Let 'em get back down Pit an'
they won't do no more runnin' around an' jumpin'."

\*  \*  \*

Union pay was not so high as wages, and wages were low.

At first the Union undertook to pay each man half his
wage. This did not go very far in families like Jim Woodall's,
where there was only one wage-earner, three children and
a wife. Or in families like the Shannons, who couldn't
manage without quiet theft and poaching even on their
full wage. Nor did the strike pay go very far in the Davies
family, where Dewi still took from five to ten shillings from
the fourteen for his own uses—drinking, gambling and
such. He didn't, however, allow any of his children to keep
money back from their strike pay, any more than he had
allowed them to keep money from their wages; and he
still expected Reenee to feed him well.

However little money Reenee was left with, whether it
was eight, or four shillings, she had to pay the rent at the
Pit Offices first—two and six. Then she had to provide
food, giving up all hope of paying any money into funeral
clubs, or of putting anything by for clothes. Nellie needed
new clothes, and so did Sarah. Sarah should have a dress
instead of going about in boy's clothes, it was indecent,
except on working days. And Jek's jacket was too tight, and
Aynoch's clogs were near to breaking. Nellie was running
naked. As for herself, she'd gone in rags for a long time now,
and she didn't expect any change.

There were many rows with Dewi over the money.
Reenee screamed and called names, nagged endlessly.

Dewi gave overwhelming displays of anger, shouting, banging, clouting his children, throwing pots and pans to the floor, and ending by slamming out of the house; or by sitting like some evil spirit on the hearth and holding all the house still and silent by his being there. Grandad Ellis almost always interfered and drew the anger of both Dewi and Reenee on to himself, and made everything worse. As for the children, from Jek down to Nellie—if they were in the house while a fight was going on, they stayed there, they didn't dare to move or speak. And after the fight had ended, but not died down, they still did not dare even to look up from the tiles.

They were losing the strike. Every night, the blacklegs went into work. They couldn't do as much as the full work-force of colliers, but they kept the Pit going, they kept the factories on short weeks instead of having to close down altogether, and the Gaffer paid them a bonus for working. It was expensive, but it weakened the strike.

Strikers and 'breakers peacefully hated one another. They lived next door and never spoke. They wouldn't let their children play together. The strikers claimed the 'Tavern', and the 'breakers had to walk to the 'Boat' across the canal for their beer. If they went into the Tavern, no one would sit by them, or stand by them; talk became quieter and they were watched—until they became so nervous that they left for the 'Boat'.

But so far no one had been hit, there had been no fights.

"Catch 'em," Timothy Russel said. "Bost their legs for 'em, set ropes for 'em, smash their windows! For Christ's sake, do summat an' let 'em know we hate 'em!"

"We don't hate them," Seth Jones said quietly and coldly. "All me life I've knowed Tommy Black. An' Macnamara."

Jim Woodall said, almost apologizing, "Look. If we're

violent, the Gaffers an' them newspapers'll be only too glad
to take it up. I know, see. I mean, I can read a bit, like, an'
I've read in them newspapers what they say about other
folk as have gone on strike, up north a bit, Derby way . . .
they'll go on about violence an' French Revolution an'
choppin' heads off an' they'll make folk think as Unions an'
violence am the same thing. Then there'll be laws, Unions
won't be allowed—they'll have the troops in for strikes, I'm
tellin' thee. Then we won't be able to do anythin'—just
like they want. So there's gotta be no violence—please.
Gotta be nothin' as'll give Union a bad name."

"A good name won't be much use if we lose, Jim."

"There'll be other strikes," Jim said, fanning away the
buzzing black flies with his cap. It was very hot, and the
stench of the muck-heaps made it impossible to breathe
deeply.

"We've got to win this strike," someone said. "We can't
have another. After all this scrapin' we *can't* lose. I'll swear
my Missis has had nothin' to eat for three days. Her ses her
has, but her ain't, I'll swear; her's been givin' her food to
the childer."

Jim said, "If we lose this time, we'll win the next."

Dewi swore. "What's the good o' that? By the time we
get enough money for another strike, if we lose this, we'll
all be dead. What's the good o' sayin' we'll win the next
one? We gotta win *this* one. Thee'm as bad as them who
say, 'Never mind that thee eat muck in this world, thee'll
have sugar on it in the next.' Makes thee bloody sick."

"We gotta have no violence," Jim repeated, like a chorus.
"Gotta keep the Union's good name."

It was very hot, and the stink grew worse day by day. Not
only their own boggin'-hole, but the stink of those for
streets around; and the pig-sties. Tempers were short, even
the babies' tempers. Every time a striker saw a blackleg, he
was irritated, as if by an intense itch, inside himself, where

he couldn't scratch. The men glared at each other and itched.

One night someone smashed the windows of every house in the Row that held a blackleg. The blacklegs wondered who it was, and had a good idea; and every time they saw him or any other striker, they itched inside. The Baden Brook was smaller every day, the stink worse, the flies thicker.

Jim Woodall apologized for the damage. He had little pride. He asked the blacklegs to join the strike, because if the strike was won, they would be taking the benefits of it without having taken any of the risk. The only answer he received was a long raspberry.

He begged the strikers to cause no more damage, whoever it was that had smashed the windows. To think of the Union's good name.

Timothy Russel swaggered up and down through the muck-heaps and the flies, and sang,

> "Join the Union while thee may,
> Don't wait until tha dyin' day,
> 'Cos that might not be far away,
> You dirty blackleg miners!"

"I'm scared," Jim said, gnawing his thumb. "I'm scared."

"Let 'em get on with it," said Anne. "Let 'em fight it out if they want to—thee can't stop 'em, thee needn't think thee can. It might be all for the best in the end."

"But it'll give the Union a bad name, Annie," he said. He had spoken that phrase so often that it was beginning to lose all sense—"Buthickle githy onion aba nem." "Somebody might get hurt, Annie."

"It won't be thy fault. Thee won't stop 'em, me lover."

Jim sat in the heat and smoke and stink and fug that was outdoors, on the doorstep that tipped forward. He listened

64

to the forge drop-hammer, thumping like his heart, and worried. He felt someone near him and looked up to see Jechonias Davies, Dewi's eldest son. Jim smiled at him, thinking idly that Jek was very like his father in build and colouring, but he had his mother's face. He was glad to be able to think about such little matters for a few minutes.

"Ado Jek."

"Ado Jim."

They said nothing else for some time, and Jim almost forgot that Jek was there. He jumped when Jek began to cough.

"Jim," Jek said, "who's right in this?"

Jim was wary. "How thee mean, Jek?"

"Are the strikers right? Me Grandad ses as we shouldn't try to alter how we are, not even in wages, 'cos he ses it's against God, and we'll burn in Hell for it." He stared fiercely down at Jim, and Jim didn't laugh.

"There's a lot of the-them as don't believe in strike say that," he said. "But Jek, it don't figure. I don't reckon so anyroad. Thee can't say as God is good, an' then say as he wants we to live like this, when there's others as have it so good."

"There's them," Jek said darkly, "as reckon that we've got no chance o' winnin', so we shouldn't starve the women an' childer, an' them as work on banks, an' we shouldn't torment them as go to work. That reckons, don't it?"

"Nobody's tormentin' the-them as go to work, Jek," Jim said evasively.

"Somebody smashed their windows, an' there's talk of doin' worse," Jek said. "There's talk o' fixin' the shaft-rope so it breaks when somebody's lowered on it—which do thee reckon's right, Jim? Strikers, or them as don't strike?"

"Strikers," Jim said simply.

"But if we can't win, like as how they reckon—then we're doin' more harm than good, ain't?"

Jim sighed. "Jek—I don't know. Why come to me? There ain't nobody could answer a question like that, not even bloody Wesley hisself. If thee want somebody to tell thee that thee'm right, an' everybody else is wrong, why not ask tha Dad?"

"He'd welt me," Jek said seriously. "If I did, he'd raise the welts on me like cart-ruts."

"Why?"

"For thinkin' summat else from what he thinks," Jek said.

Jim coughed and began to bite his thumbs. So Jim couldn't be talked to either. Jim couldn't tell him how to think, because Jim didn't know how to think himself. Jek wished he had the courage to ask Anne Woodall— she'd know, he was sure. But he couldn't even see her.

He left Jim and walked down the Row, almost having to push against the heat and the stink. Nothing to do and they were losing the strike.

The old argument began to thrash in his head. They were wrong—they were right—why?—who? Round and round and round. He was sick of it. Grandad Ellis had gone inside, so he sat on the doorstep. He didn't know that he had set his teeth together hard, and was slowly grinding them, didn't know that his hands, clenching into fists at his chin, were clawing at his face and pulling down his mouth.

Nellie came dancing up, put warm hands on his knees. "Jek, tell me about the babby—"

He slapped her across the face.

She jumped back, shocked and wide-eyed. Then her face reddened and folded as she began to bawl.

Jek was sorry. He hadn't meant it. It was like tightening and tightening a spring, pressing it further and further down. Then it is released and flies up. Jek's hand had flashed out to Nellie's face, and he hadn't even known about

66

it until she began to cry. He reached out both arms to her, partly to comfort her, and partly to shut her mouth before Reenee heard her.

But she screamed at him and ran away.

He put his head in his hands and cursed every living thing on earth. Reenee came to the door behind him, red-faced and damp with sweat. "What thee doin' to that child?"

You nag, Jek thought, why don't you drop through the floor? He said, through clenched teeth, "I hit her." It was very hot. There was a hot, sick, sad weight in his belly.

"Oh—thee hit her, did thee? Since when have thee had any business to hit her, Master?" Then, since Jek didn't answer, she brought the saucepan in her hand down on the back of his head.

Surprising even himself he reached behind and punched out. He hit Renee in the stomach, because she was lower than he was.

It was a clumsy blow and didn't hurt, but the gall of it took Reenee's breath away for a second. Then she screeched, "Jechonias Davies."

Jek got up and ran for the fields, where reasonable peace could be found.

# 5

---

Aynoch dreamed of running water, brooks and streams. In the morning his dream was broken. The hot weather had ended in a thunder storm, and although most of the lightning flashes and thunder were over when they woke up, the rain was still pouring down.

They came down into the kitchen with faces turned up to the roof, and the only sound they could hear was water— hissing against the window, trickling from the eaves, splashing in puddles around the house, plopping on the floorboards upstairs. But it was cool and they all felt more cheerful than they had for weeks, even Reenee. She had them laying out bowls, dishes, pots, pans, stewing-jars— anything—outside, to catch the water; and splashing and laughing. When Aynoch splashed her, she laughed, and then everyone else laughed harder and louder, because they had to make the most of Reenee's good moods.

They ate their breakfast to the sound of water, then waded out through the calf-deep mud to find Shanny and anybody else who would be daft enough to leave their houses. They took their shirts off before going out, for there was no point in getting wetter than you needed to; and they rolled their trousers up as far as they would go. It was difficult to run through the mud that was thickening like glue, and if they did run, then they sank deeper into the muck. The rain hammered down on them, poured from their fore-locks, down their noses, off their chins, blinded

68

them. But they collected a small work-gang, and spent an hour or so pushing the muck-heaps to the side of the Baden Brook, and trying to trample them down into a wall. They did this with every heavy rain-fall, trying to stop the brook from flooding. It was a half-hearted attempt, because they knew that it wouldn't do much to stop a flood. The muck was too soft and fluid to make a hard wall, and the water was washing lumps of it away downstream. But they'd tried, and that made you feel a little better when the water came pouring into your house, with all the filth it had collected on its way. They moved slowly and stiffly through the mud to the houses. The mud was now deep and heavy and thick. It reduced a run to a slow, stiff walk.

One by one the Davieses jumped into the kitchen, crashed against the settle, got up and squeezed around the table to the hearth, where they stood and shivered. They were chilled, soaked, and plastered with mud. A pool of water showered on to the floor.

"Get tha clothes off," Reenee said. "The way thee'm drippin' we may as well take the roof off an' have it raining in here. How am I goin' to dry them? Hang 'em over the range, an' let's hope they don't put the fire out."

So they stripped, and hung their wet clothes over the brass rod beneath the mantelpiece. The water dripped down from them and sizzled in the red of the fire. They fetched sacking from upstairs and wrapped it around themselves, sat on the hearth and felt that they were enjoying a treat, this was so warm and comfortable, and out of the ordinary routine. Reenee made them some hot water and milk mixed, a drink that tasted like rice-pudding. She grumbled all the time about their clothes, and having to dry them, but she was in a fair humour. It was raining, water was plentiful.

Money was not. Dewi, made more awkward by nagging,

69

was now taking a regular seven shillings a week from the strike-pay, leaving her seven shillings to keep house on.

Something had to go into pawn. The Highlander and his Lassie left the mantelpiece and were taken to the local pawn-shop, Fletchers. Even Jek was sorry to see them go, they'd been there so long. Fletcher gave Renee ninepence on the model, because he said these Highlanders were popular, so he could afford to be extravagant. He'd have no trouble in selling a Highlander, if Reenee couldn't buy him back. With the ninepence Reenee immediately bought a pound and a half of bacon. It had been so long since they'd eaten meat that she couldn't help herself, although she knew the money would have been better spent on more potatoes, or the ingredients for her bread-making.

Other things followed the lost Highlander: the hearth-rug, three of the chairs, one of the stewing jars, Reenee's shawl, the teapot, all of the cups except one. None of these things fetched much, except the chairs, because there was demand for them—Fletcher wouldn't give much on them because his chances of selling them were slim. But home was becoming distinctly uncomfortable.

Aynoch had started up in business, but he didn't show much profit. He hunted for fire-wood and sold it to house-wives at a farthing a bundle, they wouldn't pay more. There weren't many trees around, though, which made it difficult. Whatever Aynoch made from the selling he kept a secret, not even telling Jek, but he gave Reenee some of it.

Jek had been earning at the traditional trade for un-employed men. He waited on the canal-side, and asked the bargees if they wanted any help in legging through the tunnels. But he might spend a day in this, with only a few coppers to show for it at the end, because the bargees weren't princes either. Legging was hard work too, lying on a plank on top of a barge, and pushing the loaded boat through a narrow, low tunnel by walking your legs along the roof. He began to have ideas about something more profitable, and more in the line of work he was used to. And he didn't want too many people sharing the money with him. He went to look for Shanny.

Shanny wasn't in any of the usual leaning places outside, and he wasn't playing Little Jack Appny either. The players said that they hadn't seen any of the Shannons all morning. Jek went to their house, pushed open the door and went in.

Aunt Ruby, red-faced, was frying bacon and drinking tea from a jam-jar. The table had gone, the plates and other jars being arranged on a crate. Thomas Shannon was sitting on another crate, Daisy and Kathleen on one thigh, his arm around the neck of Liam, who leaned sleepily against his other leg. On the floor, in front of them, cross-legged and far-away, sat Francis, blinking.

"Shut the door, Jekon, shut the door," Thomas Shannon shouted. "If it ain't rainin' now, it soon will be, an' we'll be flooded out."

Jek shut the door with some difficulty, since the floor sloped away behind him, then skittered down the slope to join the family by the fire.

"Hello, Jek, my love," Aunt Ruby said. "Have thee had tha breakfast? Do thee want some bacon?"

"No thanks, Mother Ruby," Jek said politely.

"Do thee want some tea then?" As she asked the question she was dipping a jar into a large enamel bowl of tea that was kept warm on the hob.

"No thanks," Jek insisted. They must have enough trouble feeding their own, without his mouth as well.

"Oh, we ain't good enough for thee, eh? Won't have so much as a cup o' tea wi' we, eh?"

"Won't have so much as a cup o' tea wi' we!" Thomas Shannon repeated suddenly and laughed loudly.

Aunt Ruby added sugar to the black brew in the jar and

passed it to Jek. He took it, since he knew it was no use to argue, and wondered where they got the sugar from. Liam silently held out his jam-jar for a re-fill, and his mother took it, dipped it, and passed it back.

"Where's Shanny?" Jek asked, after he'd drunk half of the strong, sweet, scalding tea.

No answer for a while.

"A-bed," Liam said at last, with an effort.

"A-bed! What's he doin' a-bed at this time o' day?"

Another long pause. "Poachin'," Francis said, coughing over his tea.

Ruby turned from the pan. "They've been out all night—in this rain! They got *soaked*. I've been dryin' their clothes, I'm not kiddin' thee, I've been dryin' their clothes since three o'clock this morning."

"Catch anythin'?" Jek asked.

Both Liam and Francis gave a long, damp sigh. "No," Liam said.

"Oh," Jek said. He finished his tea, and added, "Sorry for thee. I'll go up an' see Shanny."

The staircase was even more dangerous than the one at home. At least that leant against the dividing wall—the Shannons' stairs hung from it. You had to edge your way up, trying to dig your fingers into the walls, and the steps were warped and slippery. Jek climbed up into the stair-well, so that his upper body was hidden from below, but his legs could be seen, standing on the steps. He looked in through the bedroom door—the door itself had been burned a long time ago. A cockroach scuttled towards him from the jumble on the floor. He couldn't stand cock-roaches, hated the way they reared up. He climbed the last two steps into the bedroom and stamped hard on it. It went, 'Crack', as its shell broke.

He looked for Shanny and couldn't see him. This was because palliasses and sacking covers, straw-filled pillows, old jackets, shawls and dresses and coats were humped, piled, thrown and scattered all over the floor. There were dozens of humps, any one of which might hide Shanny. Jek moved among the ragged heaps, breathing in their scent of grease, sweat and must, looking for signs of life. A gust of cold rain hit him, and startled him as he passed the window. All the glass had been knocked out, probably during the hot weather, to make the room cooler. Now the floor and the bedding under the window was damp, and would go mouldy. Why didn't some people ever think of the future?

He found Shanny—he was that very strangely shaped heap, slowly rising and falling, and snuffling to itself. Jek shook the heap vigorously, punched it, shoved it. The heap barked. Shanny rolled over, and said, "Leave off, Lee." A dog pushed its head from under the covers to sniff at Jek.

"This ain't Liam, this is Jek. Come on, Shan, wake up, come on, come on."

"What the hell do thee want?"

"I want to tell thee summat. I've had an idea."

Shanny pulled the covers down from his head. His face was paler than usual, his hair damp and tousled; his eyes were red-veined and gobby with yellow glue. An unpleasant smell of hot dog rose from him. "What?" he said. A dog that was slightly bull-terrier came from under the clothes and began to lick his face and neck. He pulled it tight against him, so that it couldn't move.

"It's Our Mother, see," Jek explained. "Her can't make ends meet. So I been thinkin' how I could make some real money—not leggin' or firewood, 'cos that don't bring in much. I was thinkin'—" he broke off as yet another terrier mongrel crawled from underneath the sacking. "For God's sake, Shan! How many dogs thee got in there?"

"All on 'em," Shan said. "They all come to me when the others went downstairs." He released the bull-terrier and it

73

raced barking round the room. He slapped the sacking, and two more dogs shot out, to join the three already up.

"Strewth!" Jek said. "Anyroad, listen. Tha know the tips, the pit tips? Tha know how Seth Jones ses as how, in the old days, they used to throw all the coal that wasn't the highest grade on to the tips, and that now they sell it all, whatever grade it is? So there'll be real good coal in the middle o' them tips. We can dig it out an' sell it. There'll be money in that."

Shanny came up on one elbow and began to pick sleep out of his eyes. He yawned again. "That'll be strike-breakin'," he said.

"No it won't," Jek said. "Will it? Will it be 'breakin''? No, it can't be ... I mean, it's for *we*, ain't? It ain't helpin' the Gaffer."

"It's sellin' coal, ain't it? An' it's people not bein' able to buy coal what makes a strike, ain't? Only that."

"No," Jek said. "That can't be right—no." Because he didn't wish it to be right. "Look—the money don't go to the Gaffer, does it? So it can't be strike-breaking."

Shanny kicked the blankets away and crawled around with his bare backside up in the air, looking for his clothes. "Thee tell that to tha Dad."

Jek thought about this. If he was right when he said it wasn't strike-breaking, Dewi would say nothing. If he was wrong—oh Christ. It would be no good trying to explain, because Dewi believed that Jek knew nothing and should never try to think, only do as he was told. There would be hell to pay.

"Well, we won't tell nobody then," he said. "We'll just go by oursen. But we'll have to go to tips after dark if we don't want anybody to know."

Shanny nodded. "It'll be dangerous, tha know, after this rain. An' we'll need summat for props. Crates or summat."

74

"Get on down here," Ruby called. "I've some bacon fried for thee, it'll be cold if thee don't hurry up."

* * *

They chose a tip outside the fence of the pit, one of two, and they crept to it when darkness was just coming on, carrying some crates and barrel hoops they had got from Isaac Walters, the Tavern's landlord. They had nothing to carry the coal in, and no tools to dig with. Shanny said the carrying of the coal would look after itself, same as everything else. The tools would have to be stolen from the Pit. They ran through the dusk to the fence, and peered in through the gap used by the women carrying dirt to the tips.

"Just think if we meet some 'breakers," Shanny said, turning his head to grin at Jek.

"If we do, thee'll laugh on t'other side of tha face," Jek said.

"No." Shanny shook his head. "It'll be all right, all peaceful, like Jim Woodall ses." He slipped through the gap and swaggered away into the dark. Jek quickly followed.

They saw no one and their confidence increased. The tool-shed was locked, but they smashed down the door and walked into the shed over the splinters. A pick for Shanny and a shovel for Jek, all neat and new. Jek protested. He wanted the spearhead job of cutting; not the follow-up job of shovelling. That was a boy's job. But Shanny only grinned and strolled away with the pick, and, since there had to be a shovel man, Jek followed.

On the sheltered side of the tip, low to the ground, they began to dig. They had to kneel, because they were used to working that way, and because the tools were designed for use in cramped conditions. Shanny used the pick first, and

75

Jek leaned on the shovel and watched his shirt tighten and crease over the changing curves and flats of his back. Then Jek moved in to shift the loose earth. His shovel was really a large scoop with two handles, and he dug it into the earth until it was filled, then dragged it away and emptied it. In the Pit, his boy would use a shovel like that to move the coal he cut. So they worked on, first Shanny, then Jek, and their shaft began to shape.

"Don't want it too big," Shanny said. "Don't want too many people to notice it." He struggled to fit a barrel hoop into the shaft to support the roof. The shaft was barely wide enough for his shoulders. A bigger one would need more digging, and more props.

"That won't hold," Jek said.

"It'll have to," Shanny told him. "It'll do for what we want."

"We'll have to make places so we can turn round—or summat," Jek said. "So we can get out."

"We can come out backwards."

"Ar—but what if we want to come out in a hurry?"

Shanny looked at him, mouth open, then squinted at the barrelhoop. "It'll hold," he said. "Don't worrit, Jek, for Gawd's sake. It'll hold."

The further they went into the tip, the harder packed the earth became, and the more rock they had to dig out and move. Work became slower and slower, hotter and hotter.

"It's workin' up to another thunder an' lightning storm," Shanny said. "Tha see."

They both had their shirts off, and were streaming with a mixture of sweat and dirt. The mud where they were kneeling was still deep and thick from the last storm, and they had sunk in it almost to their waists. Jek wondered if he would ever get his clothes dry again. They were wet from that morning.

76

It began to rain. They ignored it at first, but it came harder, drumming up the mud around them into a porridge that clung to them. "It'll weaken walls of gallery," Jek said to Shanny, and Shanny nodded, eyes squinted against the rain. Water washed over his face like a mask, streaking the clay and dirt smears on it. They had found no coal. With the rain sheeting off their backs and chilling their flesh, they walled up the tunnel with crates and earth as well as they could, the tools inside. They hoped it wouldn't flood much.

Then home. No hurry. Certainly no hurry. "I'll bet the brook's more in our kitchen by now than in the track," Jek said.

The rain pelted down, washing the warmth from them. It was full dark and they went slowly, eyes on the ground, talking quietly.

"We can sell the coal in Dudley," Jek said. "To the nailers."

"If there is any."

Jek was puzzled. "Any what? Any nailers?"

"Coal, idjit."

"Seth Jones reckons there is."

"Oh ar . . ." Shanny said. "It'll take we a month o' Sundays to get it out o' there though—Hey, Jek, when thee goin' to make it up with Ada?"

"Oh, her," Jek said, after a slight pause. "I don't reckon ever."

"We thought thee an' her was for good," Shanny said slyly. Jek could see his broken teeth, bright in the dark as he grinned.

"Thee thought wrong then," he answered sullenly. "I was fed up wi' Ada. All her little moods—spoilt blind, her was. No thought for anybody bar herself."

"Took thee long enough to find that out," Shanny observed.

77

"Thee'm just like an old woman," Jek said spitefully, to shut him up. But you couldn't hurt Shanny like that. He just grinned again. "I'm nosy, that's all."

"What I stuck with Ada for was wishful thinkin'," Jek said. "'Cos her wouldn't ever let me touch her. Her'd never leave the High Street."

"Shame," said a voice that wasn't Shanny's, and Jek jumped.

"Has he broke up with his girlie then? Aah, shame."

"Ar, ain't it a shame?"

At the curve of the tip, five men were waiting for them. From their way of standing, and their position across the path, they were not waiting to pat Jek and Shanny on the head. They stopped, facing the men, not looking at each other.

"Good evenin'. Out for a walk, are thee?" asked the centre man of the line.

Jek knew him, or rather of him. Jack Tonks from Tividale. One of the White 'Oss strike-breakers. Shanny said, "Ar. We'm out for a walk."

"Funny place to take a walk, ain't it? Round the tips? In the rain?"

"Good a place as any, Gaffer," Shanny replied evenly, politely. He sounded so calm and uncaring that Jek was filled with admiration. His own heart was stumbling along very fast, somewhere under his Adam's apple. He wanted to run, but was afraid to.

Shanny admired Jek for having the courage to look so surly. Shanny wanted to run away, but was afraid to.

"Thee'm very dirty for good little lads who've just took a constitutional round the tips."

"Ar, Gaffer," Shanny said. He smiled as an afterthought.

"What you laughin' at?" one of the men demanded. Shanny didn't answer. Another man stepped close to him

78

and clenched a big fist against his mouth. "What do you think's so funny?" the man asked, and rocked Shanny's head with his fist, just pushed his head back, threatening.

Jek's heart beat so fast it hurt. If Shanny was hit, what should he do—run or fight?

But Shanny, grave-faced, gently pushed the fist away. "Nothin', Gaffer. Not thee."

Jack Tonks was growing nervous. One thing to say, Let's give them two so-and-soes by the tips a hiding—another to do it, especially when the so-and-soes turn out to be little more than childer. "They was diggin' in the tips for coal!" he said. "You was diggin' for coal, wasn't yer?" he asked Jek.

Jek couldn't produce any sound for a second. All the things he could say shot through his head. "No, Gaffer." "Ar, Gaffer." Please, Gaffer, ar, Gaffer,"—too cowardly. "Go to hell." "Drop dead"—too brave by half. He compromised. "Wh-what's it to thee?"

"Oh, oh, What's it to me, he ses." Tonks slapped Jek's face and knocked his head back with a crack.

"Leave him be," Shanny said uncertainly. Tonks ignored him. "Now—you've been takin' coal from the tips, haven't yer?"

Jek rubbed his face and swallowed, licked his lips. That had been a taste of what was coming, and it would come whatever he said. He was afraid, he could feel teeth snapping, nose breaking—but whatever he said or did, that beating was coming. So there was no reason for soft talking.

"Go to bloody hell," he said.

Tonks punched him in the stomach, and he went down, gasping. Someone kicked him hard, and that was it. He wasn't touched again. Tonks said, "I'll teach thee to steal!" and someone laughed, but the laughter was at a distance.

79

Jek sat up painfully and couldn't see Shanny. He meant to shout for him, but "Shanny!" came out in a whisper. He couldn't breathe very well and he felt sick. Shanny came crawling out of the darkness. "Thee all right?"

"Are thee?"

"Ar," Shanny said. "Made me mouth bleed, give me a shiner, I should think, but that's about all."

"I'm all right," Jek said. "I thought we was in for a right goin' over, I did."

He heard the hiss that Shanny sometimes gave in place of a laugh. "Ar," he said. "But they didn't want to hit we, see. Maybe they did when they started waitin', but not by the time we got there."

Jek tried to get up and found it difficult. Shanny gave him a hand. They went home very slowly through the mud and rain, Jek bent over. The rain stopped before they reached the field with the pigs. They climbed over the wall and Jek parted from Shanny, walked around the end of the Row. He opened the door of the house and jumped down into about four inches of water on the kitchen tiles, crashed against the settle, picked himself up, swearing steadily in an undertone. He dragged upstairs, and thoughtfully slept on the floor, so as not to make the palliasse wet for the others. He didn't bother to undress.

In the morning, Reenee wanted to know how his face had been bruised like that. Jek glanced at Dewi, who was staring into the fire. But the old devil was listening, never mind. If Jek said anything about the digging for coal, and it *was* strike-breaking, Dewi would hear it loud enough.

"I got into a fight with Shanny," he said.

"Thee ought to have more sense at thy age," Reenee said, "And more sense than to stop out at all hours in the pouring rain an' all. Thee'm still wet! It'll bring they chest on, thee see if it don't, thee'll be coughin' an' heavin'—it

80

always happens, but it don't learn thee anythin', does it? Thee still stop out in the wet."

"It won't bring me chest on," Jek said.

"Thee wait and see," Reenee said, pointing at him with he carving knife.

# 6

---

Jek sprawled across the kitchen table amongst the bread-crumbs, occasionally picking one up on a licked finger and eating it. He didn't feel very wonderful; as Reenee had predicted, 'his chest had come on'. He had a Winter Cough. This was different from the normal, all-the-year-round cough that attacked you in times of emotion or over-exercise. The Winter Cough was chesty and weakening. It turned the voice to a wheeze, and in the mornings it made you sound like a seal barking. Everyone had them in winter, but now he suffered alone.

It was still raining. Water tumbled on to the roof, noisier even than Grandad Ellis's snores. The kitchen floor was three inches under water, soapy water, partly from the rain and partly from Reenee's washing. It splashed around their ankles when they moved, making everyone miserable and short-tempered, even the dog and cat. Keeping a fire alight was hard work, and everything was damp—chairs, clothes, bedding, food, air, everything you touched. And Reenee was drying the washing inside.

Lines were strung across the room and wet, cold clothes hung from them, and along the brass rod over the range. They filled the air of the room with a depressing dampness and a dismal, sour, soap smell. They dripped into the water on the floor with awful monotony, plop, plop, plop, plop, plop, until you wanted to bang and yell with irritation. But the worst was when the cold water dripped on to the back of your neck, or on to the crown of your head, making you jump and shiver. Or when you stood up and one or another of the icy clothes clamped itself like the grip of a shroud around your head, neck and shoulders.

Reenee, who was sploshing around through the water, doing nothing, asked, for the twelfth time that morning, "Why don't thee go out?"

"Because I don't want to go out," Jek replied, for the twelfth time that morning.

"For God's sake, stop mopin' round the house makin' me and tha Grandad miserable! Go down to the pub."

"Dad's there."

"Well, go somewhere else, then."

"It's rainin'," Jek said patiently.

"The others have found somewhere to go, for all it's rainin'! Why can't thee? Thee may as well be in the rain as in here, Heaven knows. Go an' see Shanny."

"I don't want to go out," Jek said, through his teeth, "because it is raining. I want to sit here. I ain't hurtin' thee. Why don't thee go out? See Mrs. Jones or summat."

"Don't thee lip me," Reenee said, "else thee'll get a clip around the ear-'ole, I'm warnin' thee, my lad."

She pushed her way around the kitchen, banged pans and clattered cups. She wasn't doing anything just looking for something to do. She hadn't got to cook anything, except for warming a drop of milk, perhaps, because there was nothing else in the house to eat except bread and oatmeal. She suddenly said, "Thee haven't been up Oldbury?"

He shook his head.

"Thee ain't been to Spon Lane?"

He looked up in surprise. "No! Why should I go to Spon Lane?"

"I was just checkin'," Reenee said. "They reckon they got the fever there again. Three dead already, so I've heard."

83

"I ain't been there for—weeks," Jek said.

"Last year they had the fever up Spon Lane," Reenee said. "There was one family where everyone of 'em died except the two youngest, an' a neighbour took 'em in so's they wouldn't have to go into work-house."

"When ain't they got the fever up there?" Jek asked, being drawn reluctantly into the conversation. Talk of fever and deaths suited his mood at the moment.

"It's the night air that does it," Reenee said. "Bad night air. And them bein' right on the canal like they are, it makes the air damp and bad-smellin'."

Jek flicked crumbs across the table. "Seth Jones reckons it's dirt."

"*Dirt?*"

"He means like muck an' rubbish," Jek explained throatily. "He ses as places like Spon Lane that are always dirty get more of the fever and the cholera than any place else. He ses he's noticed it, an' his father did. The dirtier a place is, the more fever it gets, 'specially in the hot weather. He reckons as the dirt gets on thy hands, an' then thee put tha hands in tha mouth, see. An' Spon Lane's dirtier than anywhere else round here, ain't it?"

Reenee pursed her mouth. "If Seth Jones spent as much time workin' as he does thinkin' up daft ideas, he'd be a rich man. Dirt! Bit o' dirt never did anybody any harm. 'Thee've got to eat a peck o' dirt afore thee'm grown.' Dirt's everywhere, ain't it?"

"Not where the Gaffer lives," Jek said. "They never get the fever out there, do 'em?"

"Oh ar, they do," Reenee said.

"Well, only every five year anyroad!" Jek shouted. A cold wet shirt fell on to his head and neck like an arctic spider, making him jump and yelp. Reenee plucked it from him and hung it back over the line. She didn't speak to him

84

again, and he slid back into his own thoughts, none of them very clear or pleasant.

Since he had received the bruise now yellowing on his face, five strike-breakers had been knocked down and kicked on their way to work, one having three of his ribs cracked; a woman had been jostled, tripped and spat upon because she was married to a strike-breaker; muck had been dropped into the bowls set outside strike-breakers' homes to catch the rain; trip-ropes had been set across their doors, for when they came out at night to go to work. Blacklegs on their way to work had been pelted with stones and muck; bricks had been thrown into the houses; and some clever lads had refused to let strike-breaking families use the boggin'-hole.

And, of course, no one spoke to the strike-breakers, not even to insult them. No one lent them things, or shared food with them, or water. The Union was refusing to help the man with the cracked ribs, even though he was, or had been, a Union member and had paid his dues every week. The Union ruled that he forfeited his membership by breaking the strike. You couldn't have your cake and eat it.

Fair, Jek allowed, but harsh. He couldn't help seeing the ruling from the point of view of the man with the three cracked ribs. That man had become a blackleg because he had five children, all under seven years old, and he had been scared of losing his job. Now, he couldn't afford a doctor's bills, he couldn't work—what the hell was he supposed to do? Because he had been kicked by men who didn't agree with his way of seeing things.

Jek found himself wondering whether he shouldn't become a blackleg. Which was ridiculous. The strikers were right, the blacklegs weren't. That was the truth—Dewi said so, Seth said so, and Jim Woodall said so.

Besides, he might as well admit it, he was scared. He had passed his friend Ayli Black the other day. Ayli was strike-breaking. His left eye was swollen and closed, looking very ugly and painful. Jek had passed by without appearing to see him. He had not asked how Ayli had got a shiner like that, and Ayli had not looked at him, or even nodded.

Some strikers were even saying that if there was an accident at the Pit—and Timothy Russel was still talking of making one—they would leave the 'breakers to rot, not make any effort to rescue them at all. This was bad. It went against all the basic laws that Jek had taken in without their being spoken. You didn't leave a man to struggle by himself, over anything. If a man forgot his snap, everyone in his work-gang gave him some of theirs. If he lost his wages, then his friends made a contribution from their own pay. If he needed money, for a christening, a funeral, a wedding, for setting up a house, or anything else, then there was a whip-round for him. The rule was as strong as any law, and the idea of breaking it was shocking.

Jek wondered if—touch wood, he was not wishing for it—if there was an accident at the Pit, just how many of the hard-hearted boasters would sit at home and not help. He didn't think many would. But even the idea was terrible. It was one of those things which should never be said, or even thought if it could be helped, for fear of tempting fate.

Shanny opened the door, jumped in from the rain, fell against the settle, got up and pushed the door shut.

"Thee ain't been up Spon Lane, have thee?" Reenee asked, straight away.

"No, Mother Reenee," Shanny said, rubbing his bruises. Rain poured down over his face from his hair and he mopped at it. "Jek—come wi' we. I got a great idea."

"He won't go," Reenee said. "He's frightened of a bit of rain."

Damn you, Jek thought. He said, "I don't want to go anywhere."

"Come on, Jek—it's important, honest."

"I ain't interested. Go jump in the cut and drown." Shanny ignored him and came to the table, leant his elbows on it and whispered, "It's about gettin' some money, like thee was sayin'. Sellin' coal."

"Oh ar?" Jek whispered back, with some interest.

"There's some old Pits in Dudley, some old shafts, nobody workin' 'em . . ."

"There is everywhere."

"Ar, I know that," Shanny hissed impatiently, "but there'll be no White 'Oss blacklegs to bother we up Dudley, will there?"

"No, but there's other blacklegs."

"So what? They don't know we."

"They soon will when they find out what we're doin', an' anyroad, there'll be no coal in an old pit," Jek said.

"There's bound to be some left, so it's worth tryin'."

"If there was any left at all, the 'brave Dudley boys' would have had it out weeks ago."

Shanny said, "Thee'm just pickin' fault!"

"All right then," Jek said, "I'm pickin' fault. I think it's a bloody stupid idea, just like thee, an' I ain't havin' anythin' to do with it. Cop me walkin' to Dudley in the pourin' rain, an' gettin' mesen killed in an old pit."

Shanny straightened up from the table and looked as if he was about to leave. Jek sat as he had sat all the morning, feet hooked around the chair-legs, head down, arms stretched out in front of him across the table-cloth. Then Shanny bent down again. He said, "Oh, come on Jek. Give we a hand. I've give thee a hand with things a dozen times, thee know that."

Jek looked at him. "All right," he said unsmilingly. "But I reckon it'll be a waste o' time, an' I ain't goin' till it's stopped rainin'. An' I'm havin' the pick, else thee can manage by theesen, an' that's flat."

\* \* \*

87

The next day there was no rain. Jek got up, dressed, and went coughing down into the kitchen, where there was Reenee, Grandad Ellis, a soggy cat, a soggy dog and four inches of water. "Where does everybody get to?" he asked.

"They'm up an' out afore thee stir," Reenee said. "What do thee want for thy breakfast? There's bread, we ain't got nothin' to put on it. Or thee can have a bowl of porridge with a drop of milk. One or t'other, make up tha mind."

Jek shook his head. "I don't want nothin'."

"Please theesen," Reenee said. "I want thee to take summat to pawn-shop for me, see what thee can get."

"What?" Jek asked. "What ain't he got already?"

Reenee shook her head. "I dunno. I'll think o' summat. There must be summat we can spare that'd raise a few bob."

Dewi came in, lowering himself carefully through the door and avoiding the settle. He seemed quite cheerful as he hauled Jek from the only chair by the scruff of his neck. Jek moved to the settle.

"Been talkin' to that Tim Russel," Dewi said, slicing bread. "He's a bright lad." He cast a glance at Jek. "He's on our side, for sure. They had a bit of a game, him and his mates, last night, from what I hear."

"Oh ar, tormentin' innocent folk again, have they been?" Reenee asked, her mouth very prim.

"They've been showin' the blacklegs where to get off," Dewi corrected her loudly.

Reenee was not intimidated. "If they hadn't got that excuse for tormentin' innocent folk as work, they'd find another. An' if thee was to give me more money instead of guzzlin' it all in beer an' running down folk that work, I'd be able to feed thee better."

Dewi looked up fiercely. "Do thee reckon that if I went back on everythin' and worked—blacklegged—do thee reckon that's decent?"

Reenee side-stepped this question, which was a little

88

too hot for her to handle. "I don't care about that—what I do care about is gettin' some more money. Thee have fourteen shillin's a week from the Union to my certain knowledge, and I'm lucky if I see seven. I want more, Old Un, I tell thee plainly."

"Well thee ain't gettin' any," Dewi said. "I'm sick o' this argument."

"Thee'll be sick o' tha meals then, an' all," Reenee snapped, but then changed her tune. "Dewi," she pleaded, "I can't manage on the money thee give me. It's worse than when thee was at work. The childer don't get above a mouthful o' food apiece, an' they all need new clothes, but I can't save for 'em. If thee could only see tha way to givin' me two shillin' more, it'd be easier. Jek—tell tha father how it is."

Jek lifted his head slightly, and saw Reenee staring at him intensely with her washed-out blue eyes. Dewi was watching him from the corners of his brown eyes, without turning his head. Jek said, "Don't drag me into it." He tried to fold up and be unnoticed. He knew that Reenee would have her own back on him for his cowardice; but he also knew that Dewi would clout him immediately if he sided with his mother. Dewi turned back to his wife and began his time-worn speech. "When I was little I had a damn sight less than a mouthful—me father spent all his money in drink an' gamblin'. We lived on me mother's earnin's an' that wasn't much. I worked in the Pit from dark to dark, from time I was four, an' I slept all day Sunday. I never saw the sun till I was twenty an' could take a day off without me father beltin' me. These childer have it easy now, they'm saft. I never had it so easy. It won't hurt 'em to do without for a bit. It'll teach 'em the value o' things. They have it easy when we'm in work, not like I had it, an' when we've won this strike, they'll have it easier. Went to school, didn't 'em? Paid for Jek there to go to school, didn't I? An' a fat lot it

89

ever learnt him. My father would have drowned me like a pup afore he'd have paid twopence to send me to school. I don't hold with little children bein' put to work, but I don't reckon this schoolin' neither. I'm all right without schoolin'. What good did it do him? He can't read, can't read a line. Well, neither can I, so what difference does this schoolin' make? Let 'em do without for a bit, I reckon, let 'em go hungry, it won't hurt 'em. Might teach the ungrateful sods to be thankful when we've won the strike. Ha! When we've won the strike I daresay they'll be wantin'— all sorts of things," he ended weakly.

"Oh ar, very nice," Reenee said. "What if thee don't win the strike?"

"We shall win the strike," Dewi said.

"Strike!" Reenee cried. "An excuse for bein' idle, that's all it is! But not just idle—want to fill tha piggish face with beer an' all, an' gamble, an' have all the pleasures—ar! An' if thee've gotta starve tha family to do it—well, that's all right! Thee can make lovely speeches to say why it's right, can't thee? Thee make me sick! Strike, my foot! Thee know damn well thee ain't got a cat in Hell's chance o' winnin'—it's just an excuse. An' I say—a damn good job thee can't win, else we'd be in a pickle. Thee'm too ignorant to tell a gentleman like the Gaffer how to do his fly up, let alone how to run his Pit!"

Dewi had stopped chewing and was staring at her, trying to stare her out. Reenee looked back defiantly, but her mouth trembled, because she realised that she had made him very angry, although she wasn't certain what, out of all she'd said, had hurt him.

Shanny pushed the door open and came in, saying, "Jek! —" Then he quietly shut the door and slid down beside Jek on the settle, trying to look as if he wasn't there.

Dewi said, "Thee'm askin' for one, Missis."

Grandad Ellis had to poke his oar in. The muscles of

Jek's arms tensed as he listened—the old fool would only make matters worse. "Thee lay off her, Dewi Davies. That's my daughter thee'm talkin' to. Keep tha tongue and th'ands to theesen."

Dewi swung round, shouting, "You can shut your trap! I keep yer, don't I? I pay for the food yer stuff down yer useless hodge, an' yer do nothin' for it! Watch it, Old Un, else yer'll find yersen pickin' on the tips for yer own livin'. You'll be out of here! An' you, Missis, what was you sayin'?"

"Thee heard," Reenee said sullenly, trying to withdraw from the fight.

"I want to be sure as I heard what I think I heard. Come on—let's all on we hear it again. Come on, come on! I want to hear yer say it again."

Jek's knuckles were waxy and white, the bone showing through his skin. He hadn't moved at all since Shanny had come in. He was as rigid as a wall. Shanny watched open-mouthed and fascinated; not because he hadn't seen similar rows in his own home, but because there was a menace under the words here. His own mother and father might rave and shout, but it soon blew over, in a day or two. Their children ignored their fights. But here—Jek was afraid, he was folded up, head on knees. Reenee was afraid—even Dewi was afraid.

"By Christ, Missis," Dewi said. "If you don't answer me in a minute I'll blind thee."

"That's it," Reenee said. "That's just like thee! Thee'm a big, bullyin' selfish *pig*! It's self, self, self with you, never a thought for anybody—now don't you dare hit me!" She backed hurriedly around the table, but came up against the chair, which stopped her retreating further. "Don't you dare hit me!"

Dewi punched her in the throat with his fisted hand, then across the face. He punched again, then changed the blow

91

to a slap, meaning to hurt her less, in a confused way. He said, in rhythm with the blows; "Don't you tell me what I am, nor what to do! Don't you give me no lip! Don't you tell me how to treat my kids. Don't call me ignorant! Don't set yoursen and your bloody God-suckin' family above *me* and my family! I'm the Gaffer in this house, and don't you forget it!"

Reenee cried. Dewi stopped hitting her. Embarrassed and frightened at what he had done he stood by, loose-jointed, not knowing what to do with himself. His helplessness made him angry again, and he looked round for someone other than Reenee to take it out on. He saw Jek and Shanny.

They sat side by side on the settle. Jek was staring with colourless eyes, in a colourless face, at Reenee. Shanny, looking sick, was staring at Dewi himself. The sight of them infuriated him. He bellowed, "What thee starin' at?" and came around the table, swinging down for the poker, a short bar of iron, on his way.

He crossed the room in two steps, but Jek and Shanny were quicker. Like rabbits they bolted from the settle in different directions; Shanny around the table, and over the chair to the hearth, Jek ducking under Dewi's arm and up the stairs. Dewi turned on Shanny. "Do you want some, Shannon? Walkin' in here like you owned the place—do you want your share?"

"No, Mister Davies, sir," Shanny said.

"Talkin' nice, ain't yer?" Dewi yelled. "I thought you and yer father was big fighters and big talkers? Or is it only big talkers? Where's all this fight gone? Am yer frightened to fight sober?"

"Ar, Mister Davies, sir," Shanny said.

Reenee said, "You leave him alone, you bloody bully, he's no relation o' yours. You leave the lad alone."

"I'd never catch him," Dewi said scornfully. "He runs

too fast." He whipped his head round to Jek, who was crouched at the top of the stairs. "You. Get down here."

Jek shook his head.

Dewi yelled, "Get down here when I tell yer to! I'll learn yer to poke yer nose into my business! I'll learn yer to jump when I speak and not to come down till I say so!"

Jek shook his head.

Dewi would never have used the poker on him if he had come down, but Jek stayed at the top of the stairs. So Dewi threw the iron bar at him. Jek ducked, and so the bar hit his head. It would only have hit his arm if he'd stayed still.

Dewi glared around the room, at crying Reenee; at Shanny, who looked away; at Grandad Ellis who was carefully watching the fire; and up at Jek, who stared straight back into his eyes and made him feel even more uneasy than he did already. He felt in his pocket, took out the first coin he found, a half-crown, and slapped it on the table. "There's your bloody money!" He stamped past the table and settle, opened the door, shouted, "An' don't wait up for me, 'cos I won't be comin' back!" Another threat he didn't mean.

When I make a threat, Jek thought, I shall always mean it. I'm not ever going to say a thing, then double back on it and give in.

Jek came down the stairs, rubbing his head. Both he and Shanny stood watching Reenee cry. Shanny knew her as a motherly aunt, always cheerful in a grim way, and he wasn't prepared for this weak side of her. He couldn't think of anything to say, and stood, red-faced, wishing he was elsewhere.

To Jek she had always been a strong thing, like a brick wall with an apron; unsympathetic, stubborn, unforgiving. She never cried. Dewi had hit her before, but she hadn't cried. Jek was helpless. He couldn't give her sympathy

93

because he had never been taught sympathy, Reenee rejected it. He couldn't help her, couldn't comfort her, he had never been taught how.

Reenee was trying hard not to cry. She kept mopping her reddening eyes with her apron and gulping; but always more tears came. Jek swallowed stickily and coughed, and said, "It's all right, Muth—" very uncertainly.

Reenee snapped, "Shut up! Mind yer own business like tha father ses. What yer standin' there for like a bloody scarecrow? Get out of me sight! I'm sick to death on yer."

Jek went around the table to the door and out. Shanny gave Reenee a nervous glance and followed him.

It had stopped raining outside. "I come to say let's go to Dudley an' have a look round," Shanny said awkwardly.

Jek walked down the Row, the quickest way to the fields and Dudley. Shanny followed. "It looks like it's going to clear up," he said, nervously pointing to the sky.

Jek said nothing.

"Hope it don't get hot again."

Jek said nothing. So Shanny shut up too. They tramped over the fields, but didn't leave the noise or the smoke behind.

"We may as well take the stuff we need with we, eh?" Shanny asked. "So if we find somewhere we got the tools an' don't have to come back for 'em?" But Jek gave him no answer.

For exploring old workings that might be filled with gas, they really needed one, or two, of the new Davy safety lamps that the Gaffer had bought for the Pit not very long ago. But these lamps were taken good care of, and numbered, each man with his own lamp. After every shift the lamps had to be taken back to the store, so that they could be counted and checked; and any man who damaged or lost his lamp was in trouble. So there was no chance of

94

getting hold of a lamp, not in the daytime with the 'guards' about. But they stopped at the tip, where they had started digging, and dug their stolen shovel and pick out. Someone else had been digging where they'd left off, and the shaft was considerably longer.

"Look at that," Shanny said.

Jek looked sullenly, but made no comment. Shanny sighed and passed him the pick. Jek took it and walked away without a word, so that Shanny had to run to catch him up.

It took them half an hour to reach Dudley, and then they tramped around the fields and rough land outside the town for the rest of the morning, and most of the afternoon. Jek was convinced that it was all a waste of time, but he said nothing. He followed from one disappointment to another without any complaint, but inside he was scalding and soon, soon, someone else was going to suffer all of that pain.

They found some shafts that obviously belonged to working pits nearby. These shafts must have been abandoned for a good reason—they were unsafe, or flooded, or gas-filled, or worked-out. And for the Coal Masters to allow them to be abandoned they must have been so unsafe that it was impossible to work in them, or completely empty of coal. That was how Shanny reasoned it anyway. Jek refused to give any help.

They found drift-shafts, which were shafts driven into the side of a hill or bank, almost horizontally—but at the last minute, as he took the candles from his pocket, Shanny's courage failed. It was very stupid to go into a pit, which might be filled with gas, or fire-damp, with a naked candle-flame. Within their own memory, men had to work at the White 'Oss Pit with no other light than a candle; and some obstinate men still used candles—but that was work, that you had to do, you had no choice. At the very last minute,

Shanny's courage ran out of him coldly. He couldn't light up a candle and stroll into that dark shaft, not to save his life.

"Can't do much, can we?" he asked, calmness and cheerfulness stretched transparently over fear. They were crouched by the drift-shaft that went back into thick blackness. A damp, close smell came from it.

Jek looked at him steadily, and saw his fear, and let Shanny know that he saw it. Shanny thought that Jek was becoming as hard-eyed and hard-mouthed as Dewi, and he didn't like Dewi. There was a viciousness about Dewi, a real delight in hurting other people with words, not only blows, that Shanny didn't understand and was afraid of. Now he saw that same viciousness looking out of Jek's face, and it made him uneasy. "What we want," he said quickly, "is a bird, and then see if we can nick one of them lanterns that close up. Must be one somewhere. We can get a bird from Dudley market. How much money thee got?"

Jek stared back. Shanny could almost hear what Jek was thinking, and it was a continual accusation of fear.

"Got none?" he asked, and laughed nervously. "Neither have I. But look—I know the man who sells the birds. Maybe he'll let we have one on tick if I talk to him. Shall we go an' see?"

Jek made no answer one way or the other.

"Well, say summat!" Shanny cried. "We can drum up some custom an' all—thee reckon?"

Still Jek said nothing.

Shanny stood up with the shovel. "I'm goin'. Thee can stop, or go, or come, or do any other damn thing thee've a mind to."

Jek watched Shanny walk away until he was small enough to seem to stand on Jek's palm, if he held his hand out. Then Jek followed.

He caught Shanny up as they came into Dudley and said

stiffly, ready to draw back if Shanny chose to be unfriendly, "Find a nailer. Ask him if he wants some coal."

Shanny eagerly responded to this softening. "Ar—that's it! The nailers, that's who we want."

Jek saw through this flattery as easily as through a sheet of glass. He suspected Shanny of sarcasm, but wasn't sure, so he kept to the cold, short way of speaking. "If we find a nailer who needs coal we can probably keep him in coal for a bit."

"That's right," Shanny agreed, and Jek was sure this time that no sarcasm was intended. He put his hand on Shanny's shoulder and said, "Tell thee what—let's go up Castle after and throw stones off Keep." Then he put both hands into his pockets and looked away, at the roofs above them. He knew that Shanny was fascinated by Dudley Castle, and it was his way of apologizing for his childish treatment of Shanny all day. The nearest he could come to an apology.

"Ar," Shanny said. "All right."

# 7

Nailers were easy to find in Dudley. The greater part of the town's population were nailers, with their own small forges, worked by their own families. But most of the nailers were still working, and resented being interrupted. Jek and Shanny peered into tiny, dark, sulphurous sheds, crammed with fires, anvils, hammers, ringing over and over again with metal being smashed on to metal; and amongst it all two or three women, men and small children, working and sweating. Jek thought that it was like peering through a key-hole into a little bit of Hell, with the fires flaring up and sparking, then dying. It coloured for him all his grandfather's stories of hellfire. Hell was a limitless nailer's forge.

The bargaining did not go well. The colliers were distrustful of the nailers, because they had been taught that nailers were the lowest of the low: dirty, thieving, cruel drunkards, who nailed the ears of boys to benches for making bad nails. Nailers had a bad reputation; the fact that a large number of them had settled in Dudley only strengthened the general opinion about them. Dudley was a wicked place.

The nailers, on their side, easily recognized their visitors for colliers because of the blue, tattoo-like scars on their faces and hands, and their very erect way of standing and elbows-out way of walking. And they had heard tales about colliers—colliers were lawless, rowdy, brutal

98

drunkards who would beat a boy with a pick-handle as soon as look at him, for no reason at all.

So dealings broke down before they started. The nailers were short of fuel, some of them very short, but it was coke they used, not coal. And they wanted to work, not waste time listening to colliers. One nailer sent a 'flash of lightning' at them as they squinted into his forge—that is, he drew a bar of iron, white-hot, from the fire, and flicked the scale that covered it at them in a shower of brilliant flakes. They staggered back, frantically brushing the burning particles from their faces, trying to shake them from the folds of their shirts. Some of the flakes burnt deeply before they could knock them away. From the forge came devilish laughter, echoing round and round.

But Shanny was always hopeful, and Jek was stubborn. They went on asking the nailers to buy coal from them, though Jek thought they would do better if they had some coal to sell. They collected the inevitable train of children with nothing better to do, who followed them, shouting, "Coal for sale! Coal for money!" and "Them from Oldbury bain't no good, chop 'em up for fire-wood and when the fire begins to crack, the fleas'll all run down their backs!" Jek suddenly turned on them and clouted heads. The children scattered and only the bravest followed them after that, racing away with shrieks every time Jek turned round, and thoroughly enjoying the game. Shanny laughed, but it made Jek very angry.

The houses in Dudley were ramshackle. Some were built back-to-back, with alleys, in an attempt to cram as many cheap houses on to a cheap piece of land as possible. For the same reason, other houses had been built around courts, with an alley to the streets. Others had been built perhaps as much as a hundred years ago, by the fathers of the people who now lived in them; and these were individual cottages, with one or two stories, one or more rooms, sprawling in whatever shape their original builder had fancied, with one or two more bits tacked on by those who came after him. But all around them crowded the courts and the alleys.

It was to one of these old, individual cottages that Jek and Shanny came next. They turned the crumbling corner of the building and almost stumbled over the family who lived in it. They were sitting against the cottage wall, in the narrow space between the cottage and the forge.

The colliers stopped with a jerk and looked down at them. Not an inspiring sight. They looked like nailers, dirty and ugly-faced. Surly and vicious. A man, three boys and two women, all grimy, sending up a concentrated smell. But one person stood out from them like a cross-eyed carpenter's thumb—a girl sitting on the far side of the group. She was neater and cleaner than the others, and she was *reading a book*. Jek watched her, surprised, while Shanny opened battle with the nailer.

"What thee want?" the nailer asked, watching them balefully, as did the women and boys. The girl looked up from her book. She reminded Jek of someone.

"We come to see if thee want any coal," Shanny began.

"Thee got some?" the nailer snapped.

"We can get some," Shanny countered.

"Thee colliers?"

"Ar, that's right; we—"

"Bloody sods," the nailer snarled. He spat at them.

Jek scowled, and said sharply, "Look—" but Shanny held out his hand to tell him to be quiet. Shanny crouched down and said, "Thee'm short o' fuel, ain't thee?"

The man allowed his face to fold into its usual expression of malevolence. Shanny took this to mean 'yes', and went on, "We can get thee coal—we can come to an agreement about the price."

"We use coke," the nailer said, and his mouth shut like a man-trap, showing teeth before the lips closed.

"But thee can use coal, can't thee?"

The nailer very reluctantly admitted this with a slight, sideways nod of the head, while he watched Shanny from the corners of his eyes.

"After all, coke's only coal, ain't it? When thee think on," Shanny continued persuasively.

Again the unwilling agreement, but the nailer added, "Coke gets hotter than coal. It's better."

"We'll get thee only the best," Shanny swore, hoping that the nailer didn't know much about coal. "What about it? Deal on?"

Jek knew who the girl reminded him of—Anne Woodall.

The nailer was wavering, but not won over by any means. He glanced around at his family, scowling, and then turned a ferocious glare on Shanny, and demanded, "What's the catch?"

"No catch," Shanny replied, face and hands open and honest.

"Got to be a catch," the nailer roared, trying to frighten him. "Thee'm on strike, no coal—then thee'm here sellin' it. Got to be a catch."

"No catch," Shanny repeated. "We'm diggin' coal for wesen, not for the Gaffers. Well—we don't want to deprive thee o' tha livelihood, do we?"

"How much?" the nailer asked.

"As much as thee'm willin' to pay, an' we'm willin' to take."

"Oh ar," the nailer said. "An' how much coal?"

"As much as we can get," Shanny said.

The nailer drew back. "This is dubious," he said.

"No," Shanny explained, "no. We get the coal, bring it to thee—and then we thrash out how much to pay for it."

The nailer brooded.

"What about it then?" Shanny asked.

The nailer was suspicious. "I ain't givin' no money till I've got the coal."

Shanny's heart shrivelled, but his face showed no sign of it. "We could do with a shilling or two, Your Honour, but we won't ask thee to give it on trust—"

"No money till I get the coal."

Shanny gave up the point. "Fair enough. Done then?"

"Ar—done," the nailer agreed slowly. They spat on their palms and shook hands solemnly. The deal was fixed. Immediately the nailer became sociable. "What would thy names be?" he asked, with elaborate politeness.

"Tom Shannon—thee call me Shan, I'm used to it."

"An' thy oppo?" the nailer asked, jerking a thumb at Jek.

"Jek Davies," Shanny said.

"Welsh?" the nailer asked with interest.

"No," Jek said shortly.

The nailer stabbed himself in the chest with his thumb. "I'm Eliakim Ansel." He indicated the boys in order of size, biggest first. "Salathiel, Harry, Josias." He pointed to the women, and said, "That's me wife Gwen and that's me wife Martha." He pointed to the neat girl and said with pride, "That's Rachel, that is, me daughter. Her can read."

Both sides nodded gravely to each other during these introductions, but then conversation died. To break the silence, Shanny asked, "Thee'm not workin' then?"

Ansel shook his head. "We've only got a bit of coke left, an' we'm all the worst for drink—we had a skin-full last night—so we let it slide. Ever seen a forge? Thee come on wi' me, my son, an' I'll show thee. I'll show thee how to make nails."

Shanny followed him into the forge willingly. He was discovering, for the umpteenth time in eighteen years, that few people are as black as they are painted, or like to paint themselves. Here was this surly nailer turning out to be as civil a gent as you could wish to meet. The others, Salathiel, Harry, Josias and the two strong women Gwen and Martha, ducked into the forge after them.

Jek stayed behind. He was much slower to make friends than Shanny, and he didn't warm to the nailer at all. Eliakim Ansel had spat at them, and had cursed them because they were on strike. It would be a long time before Jek would forget that. Besides, he was much more interested in Rachel than in the forge.

She sat on a box against the cottage wall, outlined against a high muck-heap over which fowl flapped and scratched. He leaned against the cottage corner, one clog against the wall, and watched her. But if she looked up, he looked away, and pretended that he was watching the forge door.

She was like Anne Woodall, had the same brown hair and rounded face. And there was something else about her that was like Anne, but had nothing to do with face or colouring. Her tidiness, maybe, or her—calmness. She sat peacefully.

Presently, over the clangings from all the forges round about, he said, "Thee'm enjoyin' that book ain't?"

"Mmm?" she said, lifting her head. She tucked hair back from her face.

"I say thee'm enjoyin' that book." With this opening move he left the corner and walked to her side. He bent over and tried to read the title of the book, which was set out at the top of each page. He had once been able to read; he had learnt at school, but since leaving the school to go down the Pit, he had not seen any writing, had not been called on to read anything at all—so he had forgotten how. He could pick out individual letters, but not the words. "What is it then?"

" 'King Arthur and the Knights of the Round Table'," Rachel said. Jek looked puzzled, so she explained, "They

103

was knights-in-armour and they sat round a round table."
When he still seemed confused, she skimmed through the
book until she found an elaborate, heavy black and white
picture of a fully armoured knight with very small feet,
riding a fully armoured horse with a very small head.
"That's a knight. They had all that on so's they didn't get
hurt when they was fightin'."

"He looks like a dirty night to me," Jek said. Then he
stabbed a grimy finger against the crisp, clean page. "Look
at that! A castle, like up on Castle Hill!"

"Ar," Rachel said, pulling the book away from him and
trying to rub out the mark his finger had left. "They lived
in 'em."

Jek crouched beside her. "Jockeys like him on that horse
lived in castles?"

"Ar," Rachel said.

"Lived in the Dudley Castle?"

"Ar, I reckon so. They lived in castles, anyroad."

Rachel read on in her book, but a few seconds later he
asked her, "Thee like readin', do thee?"

"Ar," Rachel said, meaning 'yes'.

"I did," Jek said, in case she should think him ignorant,
being so clever herself. "I learned to read, an' I liked it, but
I didn't read nothin' after I left school, an' that was years
ago, so I've forgot."

"I went to school," Rachel said. "The teacher took a
fancy to me, her did, an' I still go to see her now. Her lends
me books. I try an' keep 'em clean," she ended in faint
reproof.

"I didn't go often," Jek said. "Only when me mother
could spare the twopence, an' there was no jobs for me, and
I felt like it. But it wasn't a bad place. They had big red an'
yellow letters all round the room, an' pictures. They had a
picture of the Prodigal Son with the pigs, and one of Jesus
being a carpenter."

"It was a free school I went to," Rachel told him. "Thee could take money if thee'd got some an' felt inclined, an' thee could take bread an' that—but thee didn't have to."

The talk died, and Jek searched for something to say before she began to read again. "Is it good? The book—good, is it?"

"Rachel nodded. "It's good. It's interestin' like, an' excitin'—but it ain't real."

"How dost mean?"

Rachel waved her hand as she tried to think of words to express what she meant. "It ain't like livin'. It ain't like here. It ain't real. They live in castles, an' they have beds with fur blankets, an' the way they talk an' am forever ridin' through forests—it ain't real."

Jek looked up from his crouching position at the scenery she was pointing at. A strip of hard-tramped earth, filled with miserable, flea-ridden chickens; a muck and rubbish heap, sending up slow waves of decay; the rough brick, wood and tile forge, the piles of iron and rust; smoke clouding the sky and stinging in the throat; dismal washing hung out to dry, no cleaner than before it was washed because of the soot that settled on it. All black, brown, grey. No spot of colour anywhere, and the eyes grew tired. He hung his head and turned back to Rachel, nodding in agreement.

"I *could* read," Jek said again. "I learnt ever so fast. I read all the books they give me. I could read. I liked readin'. But I forgot how."

Rachel flushed a little as she realized that he was trying to impress her. "Thee could soon pick it up again," she said kindly.

Jek stared past her to the muck-heap, looking briefly into his own future. He could foretell it fairly accurately. Work, hard work, fourteen hours a day, six days a week, fifty-two weeks a year. And after that an early death, from pneumonia, or 'flu, or rock-lung, or just a bad cough. Forty years was about as long as a collier could hope to live. Most died before that, from one cause or another. "I ain't got time to pick it up again," he said.

"Thee've got Sundays. Thee've got now."

"Now I've gotta work," Jek said, "we need the money. An' Sundays—don't want to work Sundays."

"Readin's not work."

"It is when thee'm not used to it."

"I suppose so," Rachel admitted. "Anyroad—it isn't the only thing, is it? Me Dad's clever, but he can't read nor write. He can make things, he's nobody's fool."

Jek opened his mouth to agree, but then the others came out of the forge, and caught him crouching close by Rachel's side. Eliakim Ansel was delighted. "Hey up, hey up!" he said. "What's this? What have we broke up here? Come out too soon, did we?" And he winked, and nudged one of his wives, and guffawed. Ansel and his two wives looked pleased, but Jek was angry. He disliked being made fun of. Not because he considered himself too grand to be teased, but because he always thought that people were attacking him until, by long acquaintance, he had made sure they weren't.

"Come on thee," Shanny said. "We shall have to go if we're goin' to the market."

There were loud, cheerful goodbyes between the nailers and Shanny, invitations to come again were given and accepted. Jek stood back and waited. He saw no reason to be pleasant to Eliakim Ansel.

But he glanced back at Rachel as he went, to see if she was interested at all in his leaving. She raised one hand in goodbye. Her right hand. It was a round, fleshy pad at the end of her arm, thumbless, with one tiny stump of a little finger. Then he understood why she didn't work in the forge and was so neat and tidy. Before he could control his face, he felt his lip pull back in revulsion. Then he had gone.

Rachel sat down on her box with her book and gave no sign of her painful disappointment. She had liked Jek, and she thought that he liked her. But she disciplined herself carefully. That was the way things went. He'd probably thought that she was plain, and stuck-up because she could read. He'd only spoken to her to pass the time. So she was making a fool of herself in being disappointed, since there was nothing to be disappointed about. And she read in her book, all about Lancelot and Guinevere, seeming to think of nothing else.

\*   \*   \*

It was evening as they ran upwards through the streets, towards the Castle, and the market that was held below it. The market would be open for some time still, because it catered for people who worked late in factories, nailing shops and pits. But Shanny wanted to be sure of catching the man who sold the birds.

They arrived, panting, in the crowd that filled the market street. Many women had come to buy the fish that were being sold cheap at the end of the day, before they went bad altogether. The street was noisy, the stalls stumbling in and out of the traffic, the ground muddy and pooled with water. Voices bubbled all around, gossip, fast sales chatter, laughter, calling of goods. And crockery was being clattered, pigs were squealing, chickens were squawking, dogs were barking, an occasional voice rose above the hum to shout, "Blackin', best of blackin'! Ginger-bread, eight-pence a pound, plum-puddin', twopence a slice. Blackin'! Cheese, eggs, bacon, butter! Herrings goin' at a penny apiece. Must all be sold, come along ladies, penny apiece. Can't be bad! Blackin'! Best o' blackin'!" In the midst of all this terrifying noise, movement and crush, Jek lost Shanny.

His first thought was to get away from the crowd as fast as possible. But he fought the impulse, because he thought it cowardly, and instead he forced his way through the people, from stall to stall. At last he found himself looking at a pile of old and tattered books, and higher piles of fresh newspapers.

Seeing these reminded him of Rachel. Her deformed hand had shocked him for the moment, but that had been mostly surprise. He had seen worse injuries, the result of accidents in the Pit. Rachel's hand looked to be the kind of deformity people were born with, but he'd seen plenty of those too.

He thought of what she'd said about picking up reading again, and turned the pages of the books. He remembered a few of the letters and the sounds they made, but not all of them, just those that had, for some reason, stuck in his mind. 'S' was one of them, because it was shaped, and made a sound, like a snake; 'R' was another because he had always enjoyed making the 'rrr' sound against his teeth. 'T' because it was like a gallows; 'O' which was like a mother; 'I' because it was himself. Some other letters he recognized but could not remember their sound. Still, he could read enough words to encourage him—'it', 'is', 'rook', 'brook'— and he bent over the pages, mouthing patiently over each word, to see if he could spell it out. Then he turned to the newspapers, thinking they would be easier reading. They weren't. In the first paragraph he could only read 'to', 'be', and 'strike'.

It was enough to gain his entire attention. There was only one strike. He stooped low over the newspaper.

He couldn't make it out at all. Only that it was about the strike. What did it say? What was the newspaper saying about him and his people? For other people to read and talk about, but not him. If the paper was telling lies about the colliers he couldn't say a word against it. The smug, inky letters paraded before him, up and down in their lines, slyly dancing and laughing at him, because he could stare at them until his eyes rolled out and he would never be able to read what they said. It was enough to make you spit.

Rachel. She could read it to him.

But no money to buy one and take it to her.

Some part of him that didn't come from Dewi or Reenee nudged his mind, and said, Steal one.

Jek had been taught, deliberately or accidentally, that there are two kinds of stealing. There was stealing food, by scrumping, poaching or taking from stalls, gardens and counters; and that was all right. Unjust authority still punished you if you were caught, but that was authority's mistake—you knew that you had committed no crime, because you were taking necessities.

But then there was stealing things that you didn't need. You might want them very badly, but you didn't need them, and if you took them you had done wrong, you knew it, you were uneasy, and you deserved everything you got. Jek had weals across his back from Dewi's belt to prove that this was so. Jek couldn't think of more conclusive proof.

The task before him then was to establish that the newspaper was a necessity.

It was. Of course it was. It was about the strike, wasn't it?

This didn't sound very convincing, even to himself, but it was the best he could do. With the pulse jumping in his neck, Jek waited until the woman was serving someone else, then lifted the newspaper and turned to leave.

The hand holding the newspaper was taken by the wrist and lifted above his head. Jek looked into a long, thin, brown face topped with white hair. He had been caught

taking a thing he didn't need, and the scars across his back
began to itch.

"Taking my 'papers?" the brown-faced man asked.
"Have you paid for that?"

Jek expected some kind of punishment to fall on him any
second. A fist, or a strap, or some undefined, heavier
weapon. He couldn't resist because he had been trained by
punishment and worse punishment not to struggle when
Dewi beat him, for whatever reason. So it did not occur to
Jek to fight now, when it seemed that another man would
teach him right from wrong. He flinched.

But the stall-owner had never hit anyone in his life, nor
did he wish to. The defiance he saw in Jek's face he understood, but the fear made him uncomfortable. Youngsters in
this district were often treated cruelly, until they were
strong enough or brave enough to fight back. He lowered
Jek's arm but still, rather guiltily, kept a hold on him.
"Have you paid my wife for this paper?"

"No," Jek said, and then cursed himself for not having
said 'yes' as Shanny would have done. But Dewi had
trained him too well.

"That is stealing," the man said sternly. "Why did you
take it?"

"I got no money," Jek said.

"Do you always steal newspapers—or is there something
special about this one?"

"It's got about the strike in it," Jek said sullenly, looking
away across the stall.

"Oh yes, the colliers' strike. Front page news. You
wanted to read about that, did you? Well, I'll tell you what.
I'll let you read it here—I wouldn't do that for everyone,
you know."

Jek began to see, dimly, a way of escaping scot free,
newspaper and all. He wished that he had a tongue as glib
as Shanny's. "Me Dad—he—"

"Your father will want to read it too? Of course." The man spoke with approval. He was staring away into space as though he'd forgotten all about Jek, but when Jek tried to twist his arm away, the old man's hand closed around his bones with surprising strength. Then he said, "Very important subject, the strike. The stirring of the proletarian, the first realizations of his strength." Jek gaped at him, and wondered if he was somehow being sarcastic. The man smiled at his suspicious face. "You're interested in the strike? You're a believer in Equality, perhaps?"

"I'm a collier," Jek said. Didn't the old fool know a collier when he saw one?

"You are a collier? Well—take the paper, take it. If you make the news, then surely you're entitled to read it. You take that newspaper and make sure that everyone you know hears what it says. Don't you ever forget it. Er—allow me to say—that you and your colleagues are very brave men. I wish you the very best of luck. Er—tell me—are you in financial straits?"

Jek's face was as blank as a sheet of empty paper.

"Are you having trouble making ends meet?"

"When ain't we!" Jek said.

"Well, here—" The old man reached across his stall and dipped his hand into the bowl where the takings were kept. He took a florin from it, turned Jek's hand over, pressed the coin into his palm and closed his fingers over it. "Take that. It's not much, I know, but it will help, and I'm glad to be able to contribute to your strike. Good luck. But—may I give you this advice—don't steal. You'll be caught, just as you were today, but you'll find yourself in a great deal more trouble."

Jek, dazed, didn't hear much of the last speech. He stared bewildered at the man, backed away when his arm was released, then turned and pushed away through the crowd as fast as he could go.

"Not even a thank-you!" said the stall-owner's wife. Her husband stared a little sadly in the direction Jek had gone, but then he smiled quietly at her. "Never mind the thank-you, Margaret. You couldn't buy anything with it. You should be proud, woman! Don't you realize? We've just widened the horizons of life, so to speak, for that lad. It's probably the first time anybody's ever given him anything —well, now someone has. He'll think better of people from now on." But she only scowled and muttered to the books on the counter.

Jek, once out of the crowd, held the newspaper and the coin away from him, coin in one hand, newspaper in the other. His unbelievable good luck made his heart hammer. People weren't like that. People like that didn't exist. The old man should have hit him with something, or have called a policeman to do it for him. People didn't give complete strangers free newspapers and a florin into the bargain. You looked after your own, and damned every-body else. That old man hadn't been a relative, nor a White 'Oss collier . . . there must be a reason for such strange behaviour. Jek came to the conclusion that the old man had been touched in the head. Like Old Joe Wordsley—he'd given away everything he had, and everyone said that he was mad. Perhaps the stall-owner was just starting to go the same way, giving things away to complete strangers. Poor old sod.

Jek ran down through the evening and the narrow streets towards the Ansels' forge.

The streets were still filled with noise. Forges and factories were still at work, even though it was growing dark. Hammers struck, and their noise rang out in circles, becoming fainter and fainter until it died, but leaving a singing in the ears. Most of Dudley's population was coming out to begin the evening, in and around the pubs, but Jek reached the Ansels' forge without being beaten or murdered—and he was mildly surprised. From the way people talked about Dudley you'd think that no foreigner such as himself had ever come back from the town alive.

As he walked around the dark cottage he thought that there was no one in. But at the back, on her box, sat Rachel with her book.

"Thee'm here—good!" he said, suddenly rounding the corner and holding out the newspaper.

Rachel jumped in panic, then saw who it was and blushed. "What thee come back for?" she asked calmly, and tucked her deformed hand out of sight in her skirt.

Jek crouched beside her and spread the newspaper over her lap. "Will thee read that for me? I tried, but I can't make half the words out. I can't remember what half the letters mean. It's about strike."

Rachel smoothed the paper out and squinted at it. But the inky letters on grey paper were far more difficult to read in the half-light than her book. She folded the paper and put it under her arm. "Come on in the house an' I'll light a candle."

He followed her through the back door, the only door, and stood still, waiting, as her dark shape moved about in the gloom. Then she lifted a candle from the fire, the flame tiny at first, but quickly bursting into a bright yellow curve, balanced on the tallow stick. She dropped a little melted wax on to the table, and stood the candle in that; and spread the paper out beside it. Bending over, she began to read in silence.

Jek stepped away from the door to let a little more light in. He thought that the room stank even more than was ordinary and wanted to see why.

In the little light that came from the door, and the little light that the candle gave, he saw that the floor was of earth. It had been given a glaze and polish with bull's blood from the slaughter-house; blood hardened and packed an earth floor so that it could be swept, and even polished. The chimney above the ling fire was a hole in the low roof; and the only furniture in the room was the round, three-legged deal table, and some crates to sit on. Beside the fire was a three-legged pot and a tin-kettle; and in the far corner of the room, a sow and seven piglets lay snoring. But for all that, the only part of the room that wasn't spotlessly clean was the pigs' corner.

She looked up. "Come over here then an' listen." He moved to her side, and she read, " 'The Miners Continue To Be Obdurate. We—' "

"What's that mean?"

"What?" Rachel asked.

"Objurat."

"Obdurate. Er—it means stubborn, like. Pig-headed. Not givin' in."

"Thee bet we're objurat," Jek said.

Rachel continued to read. " 'We have noticed that the very miserable and unwise "strike" for an advance in wages amongst the "colliers"—' "

"Why say 'colliers' like that?" Jek demanded.

"Because it says to. See, them little marks are there to tell thee to say it like that."

"What's wrong with 'colliers'? Ain't it a good enough word for 'em?"

"I don't know. Listen . . . '. . . miserable, unwise strike for advance in wages amongst colliers—in this district is still going forward. It has so far extended to a period of several weeks, causing great deprivation and distress to the Masters and poor alike.' "

"What? What? Hey—it's a bloody shame for them Masters, ain't it? *They* didn't cause the poor no distress, did 'em? Oh no!" Jek burst out.

"Don't get so excited," Rachel soothed, "I ain't finished yet. 'The lack of the usual daily supply of coal to furnace and mill forges has become a source of serious inconvenience to the completion of contracts in manufactured iron. But we hear that a limited supply of coal has been obtained from Wales and Leicestershire, and we fervently hope that this may prevent the iron trade from being suspended. It is our opinion, and we hope our readers may agree with us, that the sooner these misguided "colliers" come to their senses and return to their work, the better, for their "strike" benefits no one, least of all themselves. They need not fear for their well-being, as there have been many evidences of the good will and care of the rich for their poorer neighbours; and we are pleased to hear from the Coal Masters that they do not intend to take action against their erring employees *even now*, provided that there is a return to work soon. We consider that the Coal Masters have done everything possible towards a peaceful, fair ending of the work-stoppage. We can only praise their tolerant and Christian attitude, while condemning the foolishness of the miners.' There thee am. That's all there is."

She looked up at him. His eyes were wide with anger, and he was speechless. He opened his mouth, shut it in a pugnacious line, opened it again, threw out his arms. "They—they—they've even got to laugh at our words!" She saw his eyes glint wetly in the candle-light. He said, "We *need* an advance in wages!" and strode, apparently without knowing it, to the other side of the room. "*They* live in big houses—with—with *gardens*—an' *servants*—an' rooms an' rooms an' rooms. They . . . they . . . *They* don't have to live all on top o' one another in one room an' all have to sleep together. It's not right! It's not bloody right!"

Rachel sat down on the earth floor beside the table and waited patiently.

"I'll bet they don't have muck-heaps outside their doors," Jek said, and then broke off, feeling in a confused way that all this was not to the point. Unable to argue, he could only cry, "We ain't misguided! We ain't unwise!" He swung round again, filled with an anger he had no way of using. "They *ain't* doin' *nothin'* to settle the strike. It's a lie! They've sat around on their fat arses since it begun, they'm just tryin' to starve we out—Christian! A—a dog's more—the Devil Hisself's more Christian than they am! They've done nothin'! There ain't we, nor the Blue Fly, nor the Fair Lady, nor the Ramrod, nor the Jubilee, nor none on 'em have heard anythin' from the Gaffers. And they ain't peaceful—the Gaffer's got right bloody manglers guardin' Pit! They—they'm always wantin' more o' we—more coal dug, more hours—but if we ask for more pay, when we *need* it—it's oh-oh, who-who'd a-thought it! Damn! Damn and bloody damn and damn and damn!"

Rachel sat on the floor, legs crossed under her skirts and listened to him helplessly swearing, because the words that he needed he didn't know, or could not bring to mind.

" 'This paper considers . . .'! 'Miserable and unwise strike . . .'! Everybody's against we, even we own kind! There's thy father spittin', an' the iron workers, even the colliers in Wales and Leicestershire! And *them*—they don't want we to have a good life, do they? They want to keep we workin'. They want little childer like our Nellie down Pit—me Dad ses there was hell of a row when some folk fetched law in to stop little childer goin' down Pit. An' thee ought to have heard Gaffer when Jim Woodall said his childer went to school—he was wicked as a wasp, Gaffer was. And thy father's on their side, not ours! I don't understand it, I don't. It's not right. It's not bloody fair."

Rachel's interest was almost clinical. She had seen her father fall into similar tempers, as fierce and fiercer, except that Eliakim Ansel never tried to give his feelings words, but only hit out—at people, not walls and pigs. Eliakim, though, grew angry because the boys made bad nails, or because his dinner wasn't ready, or because he thought that one of his wives was showing too much interest in another man. Never over something like this—Eliakim wouldn't have been bothered. If he thought about class divisions at all, it was as something he couldn't change; and anyway, it was nothing to do with him because he was an independent man with his own forge.

She waited till it seemed to her that Jek was tired, and she said, "This is all a pack of saftness. It's all talk, talk, talk. Thee can't do a thing, an' thee won't try. I've never seen or heard anything like it for saftness, and I've seen some things, I can tell thee."

Jek stopped, and watched her for a second before turning sulky. And that's it, Rachel thought, he'll hate the sight of me now.

Then he suddenly said, "They'm right, that's what makes me so bloody mad."

"Who's right?" Rachel got up and lit another candle.

"They am. In that paper. Sayin' it's a miserable an' unwise strike, and it benefits nobody, we least of all. There's nobody got any luck out on it—the blacklegs am tormented; the bank-wenches, an' the pickers an' washers am nearly all out of work whether they want to be or not— an' they get no strike pay, not from we anyroad. An' we— our family—we've got all the furniture in pawn—lost some on it—an' there's always arguments, an' me father hit me mother this mornin', an' me mother's always feelin' wicked—wickeder than her was afore strike, an' we can't open our mouths for her. There's nobody got any good out on it."

"What about if thee win an' get thy 'advance in wages'?"

"We won't win."

"Thee might," Rachel insisted.

"It ain't likely, is it?" he suddenly yelled at her. "The Gaffer—he's got all that money—he can afford to just wait. We run out o' money, we run out o' funds, 'cos we got to pay the strike pay every week, and keep the widows an' orphans and them men as was hurt in accidents as Union was keepin' afore strike. We'll run out of money soon an' we'll have to give in. Stands to sense. He'll win, the Gaffer will. We're havin' all this—this party—for nothin'. We'll have to give in in the finish."

"Thee *might* win," Rachel said. "He can't wait for ever."

"Neither can we!"

Rachel searched around her mind for something comforting to say. "Look. Thee've stayed out this long. Gaffer knows thee ain't kiddin' about strikin' now. Even if he wins, this won't have been a picnic for him, so he'll have more respect for thee in future. That's a gain, ain't it?"

There was no answer for a time. Then the blur in the darkness that was Jek shifted slightly. "Gaffer scared o' we," he said, barely above a whisper. "He'd have need to be, wouldn't he? We'll run out of funds soon an' we'll have to give in an' go back to work. But we'll have to go on keepin' widows and sick men—that'll use up what money we got left. An' then, if we want to strike again, we got to save up enough money to pay the strike pay. But that money we put in every week—it'll have to be give out again every week to the widows an' sick—dost see? We'll never save enough again. An' if we did manage to save up some— it's only need an accident—an' they ain't so far an' few between—an' bang! It's all be gone. We ain't got a chance of ever going on strike again, an' if I can think o' that, the Gaffer can. He knows he ain't got to be scared o' we."

"If thee can get funds together once, thee can do it again."

"No," Jek said. "We've just been makin' fools o' wesen— an' we're as unchristian as they am."

Rachel sighed, and sat down by the ling fire, feeding it carefully so that it wouldn't go out. "So," she said, with false aggressiveness, "thee'd pack in tha lovely strike an' give in to the Gaffer right now, if thee could?" That ought to bring results, she thought.

There was another long pause before he answered. "No. Well—see—*I* wouldn't. But it ain't only me. I'd stick to it, I'd stick it out as long as the Gaffer could an' longer, I would. But there's the top-side workers. They ain't got no strike pay as I know to, but they'm out of work, most of 'em, because of the strike. An' there's me mother, worried to death, an' Our Nellie, hungry all the while, an' Annie Woodall throwed out of her house. An' the blacklegs—well, they're only doin' what they think's right."

"So," Rachel said, "thee'd pack in the strike for them? Thee'd give up all hope o' winnin' an' gettin' higher wage for blacklegs and childer an' top-side workers?"

Another silence. "I think we ought to," Jek said quietly.

"I think," Rachel said, "as thee should make up thy mind whose side thee'm on."

"Eh?" Jek gasped.

"Am thee on strikers' side or blacklegs'?"

"Strikers!"

"Thee think strikers am right then?"

"Ar! We *need*—"

"Thee'm *sure* they'm right?" Rachel asked.

"Ar!"

"Well then, why worry about blacklegs an' pickers an' washers? Thee think thy side's right—then stick to it, an' don't give a damn for anybody else!"

"But—but—" Jek crouched down to explain. "It's not fair, it ain't fair on 'em."

"That's hard luck," Rachel said, keeping to her hard-hearted role. "Look—thee'm on strike's side. The blacklegs chose to be blacklegs. They knowed what they could expect. They ain't worryin' about thee, bin 'em?"

"Some of 'em might be," Jek said.

"Oh shut up. They ain't givin' a tinker's damn about

119

thee—why should they? They got more sense. They got their own problems an' they stick to 'em, not like some folk I could mention. Ain't I right in sayin' that if thee win, the blacklegs'll get the pay-rise as well as thee?"

"Ar . . ."

"There thee am! Walkin' away with the money without none of the risk! People like that ain't worth worryin' about. I'm right, ain't I?"

"I suppose so," Jek said guardedly. "But—"

Rachel jumped in before he could finish. "The pickers an' washers—why don't thee get 'em into this Union o' thine? Thee'd have more people to collect money from for the funds. What's the matter with 'em anyroad? Why ain't they in the Union already?"

"It was just colliers," Jek mumbled, "men underground."

"Why?"

"I don't know, do I? I didn't start it!"

"Well, thee can change it," Rachel needled. "Thee can get 'em into Union an' then thee won't have to worry about 'em, will thee?"

Jek said nothing. All the forges had stopped now and it was full dark. There was only muffled laughter and shouting from the streets.

"All that," Jek said at last. "That's all right, what thee said, if thee can stick to it an' never think about other folk at all—but what about th'own family? They'm starvin'—childer on thy own side. There's sick men got to live on less than we, an' widows with families. And there's a babby dead in Brades Village, so we heard a week or so ago. What about them then, eh?"

Rachel sighed. "Thee've just got to stick—thee've just got to do what thee can and carry on, ain't thee? It's no good sayin', 'Ah, we'll have a strike an' get some more pay'—an' then givin' in when a babby gets sick, is there?"

"When a babby dies," Jek said flatly. "There's plenty sick."

"Well, it's still no good givin' up—thee've just got to go on, that's what I think, anyroad. I mean, if thee win, thee can feed the childer up again, can't thee?"

"It sounds bloody hard to me!" Jek said. "It's all right for thee, ain't it? It's nothin' to do with thee! Thee can sit there an' tell me what to do—but thee'm all right! Thee'm out on it!"

"I was only tryin' to help," Rachel said.

Jek hung his head and grinned. "Well—it does help. If I could just hang on to what thee said. I could tell Jim Woodall about takin' pickers an' that into Union—he'd see it, I bet. He's clever, Jim is. But the rest—I think—I think I should have to stop away—away from everybody that's in the strike, like, afore I could get any—comfort—from that." He blushed at the personal nature of what he'd just said, and wished that he could take it back inside himself and hide it, but Rachel was flattered.

Jek said, "Jim Woodall's our Union Man. He's Annie Woodall's husband. Her's nice, Annie is." He looked at his fingers and muttered, "Thee look like her."

Rachel laughed nervously. "Thanks for nothin'! I'll bet her's forty if her's a day!"

"Her's nice," Jek growled. There was laughter out in the street, making the room much quieter. "Her's pretty," Jek managed to get out, and then looked up abruptly. "Where's everybody gone then?"

"Out to the Pub."

"Will thee be all right on th'own until they get back?"

"Ar, 'course. I stop on me own often when I've got a book I want to read."

There was more shouting outside. "Only I could stop— until they come back . . . if thee like," Jek stammered.

121

"Thee can stop if thee want to, but I shall be quite all right on me own."

Jek took that to mean that she couldn't wait to get rid of him. It was the 'quite' that did it. "All right," he said, getting up. "If thee'm sure. Thanks for readin' the paper. I'm sorry," he said through gritted teeth, "for ravin' an' shoutin'. I got carried away like." He thought of giving her the florin, but held on to his sanity. The Davieses needed that. At the door he looked back. "Be seein' thee then, shall I?"

"If thee like," Rachel said coolly.

She thinks I'm saft. "Well—I might then. Tara. Thanks." He walked around the cottage and into the noisy streets. He felt exhausted, and strangely near to crying; he wondered how he could finish the seemingly endless journey across totally dark fields to home. Inside his head was a confused welter of thoughts and he couldn't pin any of them down. He concentrated on putting one foot in front of the other.

Rachel, in the dimly-lit cottage, got up and went to the table, where she folded the newspaper neatly, then took it and laid it under the straw and rags that she slept on in the next room. She would read it all when she had the chance, not just the part about the strike. She came back, blew out the candles and sat in the dark, waiting for her family to come back. She shivered suddenly, for no reason, and looked nervously around. Someone walking over her grave—bad luck. She reached out and tapped the wooden table three times, one for God, one for Wod and one for Lok.

Jek, tired and empty-feeling as he trudged across the dark and dangerous fields, suddenly shuddered for no reason. Someone crossing his grave. Bad luck. He stopped and looked around, drew a hand across hot, heavy eyes. Maybe it was a warning about the old pit-shafts, clay-pits, even quagmires, into which he might stumble. He lifted a leg and touched his wooden clog, then went on more carefully.

He reached home safely, very cold, very tired, feeling a little dazed. He couldn't remember how he got there; he supposed he must have walked in the right direction, but he couldn't remember choosing that direction or noticing anything along the way. It was as if he had fallen asleep somewhere out in the fields, and had only woken outside the Row.

As he walked up the Row he heard a child crying with long screaming wails. He opened the door of the house, fell in on top of the settle and got up to shut the door without knowing what he was doing, or that he was bruised. The child was crying upstairs in the Davieses' house.

He climbed up the stairs on all fours and the cries grew louder. He opened the bedroom door and squeezed in. Everyone except Reenee and Nellie were asleep, or trying to sleep. Nellie was crying, and Reenee was shaking her violently, so that her head flopped, and hissing, "Shut up, shut up, shut up, you brat, shut up."

Jek watched with a sick stomach. He said, "Muth; Muth—don't—leave her be—"

Reenee glared at him. "All right for thee!" she said. "Thee ain't got no work to do, have thee? It don't matter whether thee get any sleep! Well, I need my sleep an' this brat ain't goin' to stop me!" Reenee was nearly in tears herself. Still, she stopped shaking Nellie and hugged the child to her, rocking her. She saw Jek still staring and hissed, "Get to bed, else I'll gi' thee a taste o' me hand!"

Jek didn't move. He was wondering how long it would be before Reenee lost her temper again. He could understand her anger, but that didn't make it any better for Nellie. "Her's hungry, Muth, that's what it is."

"I know that! I ain't stupid, though thee might think it when I let mesen get saddled with house-full of ungrateful

123

brats! Well, her'll have to stop hungry, 'cos there's nothin' in the house. Shut up Nellie, now *shut up*!"

Jek said, "Shall I have her? Give thee a rest?"

"Now what do thee know about childer? Get out of me sight, you bloody useless article!"

But when Jek stooped to take Nellie from her, she let him. Jek boosted Nellie high against his shoulder and snuggled her down, rubbing his chin on the crown of her head. Nellie snuffled, and then stopped crying. Jek wasn't fooled. He knew that she had stopped crying to spite Reenee. He carried her behind the curtain and got under the coverings with her, next to Aynoch. He didn't bother to undress, he was too tired. Curled around her, he fell asleep quickly, as worn out as Nellie was herself.

# 8

---

Breakfast. Grey light filtered through the fresh, rain-laid mud on the window, silvering everyone inside. The rain seemed to have stopped, for the day at any rate, and Reenee had risen early and baled the kitchen out, so that the floor was fairly dry. It seemed strange for there to be no water on the floor—you got used to such discomforts so quickly.

Dewi sat at the table, gloomily eating bread, his malevolent gaze on his eldest son. He suddenly demanded, "What time did thee get in last night?"

Everyone jumped and looked round guiltily to see whom he was talking to. "Who, me?" Jek asked, when he saw Dewi and everyone else staring at him.

"Thee," Dewi said.

"Late," Jek admitted.

"Ar—I don't need to be a bloody prophet to know it was late, do I? Where had thee been till that time?"

"I went wi' Shanny," Jek said.

"*He* come home hours afore thee," Dewi told him. "*He* said he'd lost thee—in Dudley. Now where did thee get to?"

Jek tried to stare him out and couldn't. He looked down at his hands and said, "I went to see this girl."

"What?" Dewi shouted.

"I met this girl—"

Dewi got up. "I've told thee about goin' up Dudley, my lad. If Shannon wants to let his tribe run wild, that's his business, not mine—but I've told thee—"

Jek stood up too. "I went to see her," he yelled above his father's voice, " 'cos her can read, an' I wanted her to read me a piece out a paper I—I found."

"Oh ar?"

"Oh ar! It was about the strike, an' I wanted her to read it—it said—said as we're unwise an' misguided, an' that strike was good for nobody, least of all we—an' that they was bringin' coal in from Wales!"

"What?" Dewi repeated.

"Bringin' coal in—from Leicestershire as well."

Dewi sagged. "Well, that's done it. That has bloody done it."

Jek sat down again, seeing that Dewi had lost interest in him. Dewi absently sat down too, and his face gradually lost its truculent expression. He looked almost friendly.

No one spoke for a while, and the house creaked to itself around them. They were all tired and a little miserable, like the weather. And they were all hungry. Nellie had fallen asleep under the table.

Reenee leaned against the wall, folded her arms and said, "The sooner thee'm back at work, the better."

Dewi gave her an evil look, but said nothing in reply. Her face was swollen and bruised from the punches he'd given her. He rubbed his own face, and then said, not looking at her, "I was talkin' to Methody Bates, an' he ses as we're going to have to see about lowerin' the strike pay. Ses we'm going to vote on it."

Reenee sighed. Jek jumped down from the stairs, and laid his florin on the table. Hearing Dewi talking about money had made him remember it.

Both Dewi and Reenee looked at the coin, then sharply up at Jek. "Where d'thee get that?" Dewi asked suspiciously.

"I found it," Jek said.

"Thee found a florin?"

"Ar. In the gutter." That was Jek's story, and he was sticking to it.

"Ar; well; folk go around droppin' florins, don't they?"

"Somebody must have dropped it," Jek said stubbornly, "because I found it in the gutter."

"We'll believe thee, thousands wouldn't," Reenee said, snatching up the coin. He knew Dewi didn't believe him. But if he told the truth, Dewi wouldn't believe that either. Anyway, it didn't matter, because Dewi wouldn't say anything—they needed the money. "I'm goin' out," Jek said. He didn't want to be glared at all day. Made him uneasy, as if Dewi was putting the evil eye on him.

Dewi made no objection to his going out, so Jek edged around the table, away from him. He climbed out of the door and tramped off down the Row, through the mud. Shanny was tramping up the Row. With a toothy grin he held up a Davy-lamp. Hands in pockets, Jek ran to him. "How d'thee get that? I thought we was goin' to use a bird."

"Bird-man had gone when we got to market," Shanny explained, "an' now we got this. Thee know how we saw as somebody'd been diggin' in the tip where we started? Well, I give 'em hand when I come back from Dudley last night. There's five of 'em an' they been diggin' all hours since we left off. An' since we started it, they give me some coal an' lent me a Davy-lamp they'd nicked. There's stacks o' coal in that tip—they must have been wasteful sods in old days."

"That's good," Jek said with one of his rare grins, and he fell in beside Shanny as they walked around the end of the Row.

"I don't know as we wouldn't do better to dig in the tips now we know there's plenty o' coal in 'em," Shanny said.

There was a pause, and then Jek said quickly, "No, let's go to Dudley."

"Hey, thee've changed tha tune, ain't?" Shanny said. "Thee'm the one as didn't want to go to Dudley."

"Ar, but we got the lamp now, an' we spent all that time trampin' round Dudley. We may as well go to Dudley now—we've left the tools there. An' them other men am diggin' where we started."

Shanny puzzled over this change of heart as he stepped on to the field wall, ran along it and dropped off into the long wet grass. He swung round and grinned at Jek. "It ain't that wench? That nailer's wench?"

Jek shoved him off balance, making him stumble, and said, "Let's go if we're goin'."

"It is! It is!" Shanny said, more to tease than because he believed it. "It's that nailer's wench! Fancy fancyin' *her*."

Jek immediately thought of Rachel's deformed hand, and anger filled him. "What wrong with her, you rodney?"

"Oh," Shanny said, still teasing, but with less pleasure, "he *does* fancy her." He walked backwards in front of Jek. "Well, her ain't bad-lookin', I'll say that, but her ain't the kind *I* like." He spoke as one with much experience of women and their looks.

"Ar; well; we all know the kind *thee* like," Jek retorted.

Shanny felt that this, in some dark way, was a deep slight. He grinned maliciously, his eyes becoming slanting black slits. "I know why thee like that nailer wench," he said. "It's 'cos her looks like Annie Woodall, ain't it? I've seen thee moonin' after Annie Woodall."

Jek was speechless with surprise for a second. He had thought his admiration for Annie Woodall very secret, but Shanny had seen it; and he'd also thought his likening of Rachel to Annie private, in his own head—but Shanny had seen that too. "Why don't thee mind thy own bloody business?"

"Thee goin' to make me?" Shanny asked, walking

backwards, grinning insolently. But he stopped in surprise as the Baden Bull, or hooter, sounded. Both their heads jerked round to stare in the direction of the Tool Factory.

The six o'clock had blown, and the bull shouldn't sound again until the break at twelve. Anyway, it didn't sound like the blasts given for the time. It sounded as if someone was pulling with both hands, urgently, to open the valve and let the steam through.

Three whistles in quick succession blared out across the fields and houses, making the pigs squeal. Another blast followed, and Jek looked quickly to Shanny, who stared back, tense and fearful.

Three more blasts came, loud, fierce bellows, and then a fourth. Eight blasts. They gaped at one another in horror. Eight blasts. The signal blown on the White 'Oss bull to tell everyone within earshot that there had been an accident.

"Accident," Shanny said shrilly and jumped forward. But Jek caught his arm, swinging him round. "Hang on— why's the Baden blowin' it? That's the Baden, not the White 'Oss. What's goin' on?"

Shanny dragged him forward. "It's a bloody accident!" he said. He broke free and began running for the White 'Oss. The White 'Oss signal was the White 'Oss signal, even if blown by the 'Palace' orchestra, and when you heard it, you ran to the White 'Oss Pit.

Men came pouring from White 'Oss Row, women too. Jek ran back, bounded over the wall and jostled into the people standing outside the Row, asking, "What's happened? What's happened?"

Nobody knew.

Then a man came out of the Baden's tunnel gate. He was a White 'Oss collier, from Baden Village. He was dirty, bruised and exhausted. Jim Woodall saw him and ran across the road to him.

The collier explained, almost in a whisper, and the people passed his words back.

He and four of his friends had been digging in the tip for coal. They had been working hard and were tired. To save themselves time they had made the galleries very narrow, had used few props, and those made of crates and such. The galleries had begun to fall in, and they'd had to get out; but they couldn't move faster than a very slow crawl, because the galleries were only just wide enough for their shoulders. He'd been last in, first out, he'd made it. The tip had fallen on the others. For a time, dazed, he'd tried to dig them out himself, but he finally realized that he had no chance of managing alone, and ran to the Pit for help. But the Gaffer's 'guards' turned him away. They wouldn't let him explain. So he'd run two miles to the Baden, gone in and asked for the bull to be sounded. He leaned heavily against the wall of the Baden.

Jim Woodall looked for his wife, whom he knew he could trust. "Annie! When they come past from the Fair Lady— or anywhere—tell 'em what's happened and where to go!"

And then he ran after the other men. And the women ran for the blankets and beer, and old rags, before they followed the men.

# 9

---

The guards wouldn't let them in. There'd been no accident. They had orders to keep all strikers away, and they would. They'd heard the bull go, ar, but there was no accident. Now then—were they going to go?

Someone pointed out the people coming from White 'Oss Row. A long tattered streamer of men and women, some lagging and panting, some hopping, some still running strong. At any other time it would have been a funny sight. Some of the guards laughed.

These newcomers didn't head for the pit at all; they made for the tips and beckoned those at the Pit gates to follow.

They found the tip that had slipped. Its shape was altered, and reaching from the black earth was a hand. The hands that reached from behind curtains in melodramas at the Palace shot into Jek's mind—but this was real. Not funny, or exciting, just real.

They began to dig. Several of them tried to dig at once and got in each other's way, until Jim Woodall called out the names of those who were to stand back. They dug with their hands, because the man was so close to the surface that they were afraid to use tools in case they injured him further.

The three men dug as fast as they could, and then tried to dig faster. They clawed out the earth in handfuls, sweating and panting. It had been loosened by the slip, but it was sticky with rain; it had fallen in heavy clods, hiding pieces of stone and coal, some of them huge lumps which had to be moved by two men or more. When a man stopped, without knowing he had stopped, and tried to wipe his face free of sweat, Jim Woodall jerked him backwards, and took his place. There were men all around who were only too ready to take the places of those who worked, and every time a man slackened a little, he was replaced. The work barely paused.

Shanny scrambled up the side of the tip and began to swing a pick-axe, in line with the diggers below him. Jek and Ayli Black ran and crawled up to help him, with another pick, and hands. There were four other men buried, and the sooner they could be reached, the more likely they were to be rescued alive. But the loosened earth from their digging fell down to cover the digging below, no matter how they tried to stop it. They sent dirt and stones tumbling down with every movement of their feet; and their picks split the crust of the tip. The diggers shouted angrily, and Jim Woodall's voice came up, "Get down from there! Thee bloody idjits, thee'll have the tip slippin' again!" They stopped, sweating, gasping, wiping their faces and smearing them with dirt.

The man's head and shoulders were uncovered. No one knew how long they had been working. They couldn't tell whether the man was alive or not, he was lying on his face. They couldn't tell if he was hurt, because of the mud that clung to him.

Jek slid and fell down to the bottom of the tip, where he began to claw out a new tunnel, one that would run in at an angle to those buried behind the first man. He was joined by Ayli Black, who said, "Mind thee hands, I've got a pick." Jek sat back on his heels as Ayli's pick slammed in. His hands were cramped and sore, tingling. He looked up at Ayli and something bothered him. He said, "Thee'm a blackleg." Ayli gave no reply. Others came to help, finding a rhythm so that they didn't get in each other's way. Jek cleared the loose earth, with three other men. When they weren't needed they watched the diggers, waiting for a chance to use a pick.

The first man was uncovered. There was an outbreak of chatter to let the other diggers know, but that was all. None of the digging stopped. There were three other men to be dug out. "He's alive!" someone shouted. No one said: For how long? but the question was buzzing all around them, and no man looked at another.

Jek had to stop. The ache in his arms was unbearable, through trying to dig so fast. His chest hurt and he felt a little dizzy. Just one good breath, just pause enough for one good breath . . . but as his pick stopped a man took the handle. It was a Blue Fly collier. He took the pick and began work fiercely where Jek had left off.

Jek shuffled backwards, out of the way, and sat down. It was some time before his heart stopped thumping and he was breathing normally. He looked at the digging, and it seemed to him that they had cleared a lot; in the next second that they had only scratched the surface of all the earth they must shift to recover the three buried men.

Most of the men now digging were 'foreign' colliers, from the 'Blue Fly', the 'Fair Lady', the 'Red Lion' and the 'Ramrod'. They had come when they heard the bull blow eight times, and had taken the places of the tired White 'Oss men. Then he noticed that the colliers digging and waiting to dig were blacklegs as well as strikers. He was tiredly surprised to see members of the Gaffer's guards sweating alongside the colliers too. He got up and went to see the one man that had been dug out.

Women were cleaning his face, and trying to see if his bones were broken, how many and which ones. Jek came up behind them and, almost unwillingly, looked down. He only blinked once. The man wasn't nearly so horrible as the victims of some Pit accidents Jek had seen. His nose had been flattened, his mouth split. A little spout and bubble of blood kept springing out from between his lips, and running over his face, down his chin and up over his eyelids. Jek turned away.

They were uncovering the second man, but he was dead. His head had been smashed in. Jek saw a little red of it, but turned his head away before he saw more. It was cowardly, but he couldn't make himself look, even so. He hurried back to the second digging, to see if he could be of use. He relieved one of the Gaffer's guards and got down on his knees with a shovel. There was no hope among the diggers that the men in the tip would be alive, or that the first man would live; but they kept on. More diggings were started, because of the nagging fear that the men might be alive, but would die if they weren't rescued soon. The diggers drove themselves as hard as they could, the sweat pouring over them. This minute the men might be alive; the next minute they might be dead.

Jek pushed the shovel hard into the earth loosened by another man and jarred every bone, every joint of his arm, from wrist to shoulder. He wrenched the filled shovel away and twisted over to empty it, stretching the muscles along his arm and down his back. But just as hard he drove the shovel in again, the sweat flowing as freely as stream water down his brow and lip, down his chest and sides, and back and thighs, soaking his scalp. He jerked the shovel away and, gasping, turned it over; and he repeated, sometimes aloud, sometimes not, "Please let him live, please let them be alive, please, please let them be alive, let him live, please, please, *please* let them be alive. . . ."

A man took the shovel from him and he crawled away. He saw it was late afternoon. It couldn't be that late already? Yet the morning when he had heard the bull blow seemed three lifetimes away. He tilted his head back, face dripping, to see the yellow sky, dragged in a deep breath, doubled up to ease a stitch in his side. His gut was sharp with hunger. Water for breakfast, and then all this work. When he straightened he could see nothing but grey and black sparks, flashing in the air in front of him. They cleared, and he went to the first digging. He found that he couldn't control the direction he travelled in, he staggered from side to side and his head seemed a long way from his feet, connected only by the thinnest of threads. And when he reached the digging, it was a second or two before he could understand what he saw there.

They had pulled the third man clear. He was dead. And the first man had died too. Jek saw Shanny sitting away from the work, hugging his arm to his chest. But then he saw a digger throw up his head for air, and took the man's place.

For a second, as he went to his knees, he wondered whether it was sensible to take the pick. He ought to let someone who was less tired dig, if there was any such person left. But digging, however tired he was, was better than being idle while others worked. He threw himself into the work with as much energy as he could find. It wasn't much. His chest hurt, every line of his body ached at the effort. He raised a hand unconsciously to wipe his face. A man pulled him out of the way and fell to his knees in Jek's place, began to dig. A blackleg from Baden Village.

Jek crawled away. At a distance he collapsed and lay face down. He felt as limp as a rag-doll. He hadn't been so tired since the first terrible months down the Pit when he was ten. He felt that he could go to sleep now, and with any luck never wake up again. At that he was ashamed and heaved himself up. Everyone was feeling like that, but they weren't lying down and oozing self-pity. He got up and dragged back to the digging, waiting for a chance to dig again, with a crowd of others. He saw his chance, but was pushed aside by a guard, who said, "I'll be more use than thee," in passing.

They found the head of the fourth man. He had been buried much deeper than the others, he must have lagged behind, or perhaps he had been hurt in the running and couldn't keep up. Jek saw some men from the Baden Tool Factory who had come to help. Surely it wasn't so late that work had stopped at the factory? But faces and figures were becoming obscure in the evening light. There seemed to be no time, only this present, digging for buried men. This was all he could remember, ever, and it was all he could expect of the future. There would be no tomorrow, only this deep, nightmarish yellow light with crowded, shifting, boiling figures, digging dead men from the earth, like rooting up potatoes. Jek was cold with horror and went floundering through the crowd, looking for a chance to dig and end this, dig all the men out and end it.

He was working again, dragging the earth aside, digging as fast as a dog does, then becoming slower and slower. He just couldn't keep up that pace any more. He was working in a state between sleeping and waking, he saw and heard nothing. Someone pulled him out of the way, put him to one side like a tool that isn't being used, and gave him a push. He walked obediently in the direction he was shoved, until the crowd stopped him, and then he turned and stared blankly at the digging men.

The fourth man was pulled free and turned over. Men crouched over him and brushed dirt from his face. Fingers felt at the throat for the pulse, hands were held over the mouth for breath; a man laid his head on the body's chest, listening for the heart beat. He raised his head sharply, his mouth shaping the word, 'Dead' and looked directly into the widow's face. His voice died before he spoke, and he turned his head away. Another man said, "He's gone, mother," which he thought was a more acceptable way of putting it.

The four bodies were arranged neatly in a row, like beads on a string, and people stood numbly looking at them. Jek wasn't very sure of anything that was going on. He was drifting further and further away inside himself, he saw things in dim blues and greys. He recognized people and objects, but didn't feel connected with them. He felt as he would if told of an accident in Scotland—sorry, but not concerned. He saw Dewi scowling, and Shanny still nursing his arm, and Reenee fussing over tired Aynoch; but he didn't feel that he knew these people, only that he had seen them somewhere before.

Doors were being brought from the White 'Oss, un-screwed or broken from the various sheds. The bodies were laid on them, and the brothers, sons and work-mates of the dead men pushed forward for the honour of carrying them home.

The colliers stayed behind, standing around the heaps of earth they had made with their digging while the sky darkened above and around them. It seemed disrespectful to the dead to leave and go home, but there was nothing else to do. Slowly the crowd broke up and drifted away in silence.

Jek was home but didn't know how he came to be there. He fell in at the door and staggered across the room, steadying himself on the furniture. The mice and cock-roaches that had come out in the family's absence scuttled away from him, but he ignored them.

The stairs led up to enormous heights, into darkness, far higher than he could climb, it was impossible.

He climbed them.

He leaned on the bedroom door until it opened.

He stumbled along the wall until he felt the curtain against his face.

He went to his knees, then sprawled on to his stomach and slept.

*     *     *

Jim Woodall sat in the little church and stared up at the crucifix over the altar, as he waited for the curate, or the Vicar, or whoever. He shifted uncomfortably on his seat. He felt out of place in this dusky, quiet building. He wouldn't have come near it if he hadn't had to. There must be some way of contacting the clergy that didn't mean sitting about in churches, but Jim didn't know of them. He didn't know much about clergy at all. So he had come to the place where, he reasoned, one or another of them must come sooner or later.

He sighed and glumly admired the tombs set into the floor. Why did they put tombs in the floor where you had no choice but to walk over them? He looked at the altar, counted the steps up to it, gaped at the high roof. All to avoid his own thoughts. He didn't want to think that the strike was over and done with, and they had lost. The Union Funds, already reduced through the payment of strike pay, now had to support four new widows and their families, and pay for four funerals. Six pounds each: hearse, two carriages, coffin, the lot. A lot of money. He would have tried to persuade the undertaker to provide only two carriages, and drop the six others, and so pay for the funeral more cheaply—but people set great store by a good funeral, a big send-off. He couldn't swindle the dead men and their widows out of a good funeral. So twenty-four pounds—he blinked at the thought of it—would have to be found from somewhere. From the funds, from a whip-round, from collecting door-to-door.

Then there was the wake. The food and the drink, the

candles to watch the body by—normally the families would pay for this themselves, with the help of a whip-round; but, because of the strike, savings had been spent, there was very little money anywhere to whip up. The Union would have to pay.

Then there were all the other widows from all the other accidents; all the men in those accidents that hadn't been killed, but were too badly injured to work; all the men who were sick, or had wives, or children who were sick and couldn't pay the doctor (a broken leg, set, could take eight years to pay off, a little each week)—the Union had still to pay out to these. Jim hoped it could be managed. One thing was certain. The Union couldn't afford to pay strike pay any more. Methody Bates had said so, and Methody was a clever man.

There was a stained-glass window, but the sun wasn't behind it and neither the picture nor the colours showed up well. But there was some central figure dressed in blue. The reds, greens and blues around were dark and the edges in black. He shivered, and glanced apprehensively around the dark building. Cold stone and old, black wood. Churches were always chill, unwelcoming places. He associated them with death and wished that he'd never offered to make the funeral arrangements. But—people asked him to, and he was the Union Man. He couldn't refuse.

He sighed and rubbed his hands together. The air seemed to have a weight as heavy as the stone which enclosed it, and to bear down on him. The strike was over. Soon a vote would have to be held, to decide whether or not to end it. Only a formality, as they said; he had explained the situation and nearly everyone would vote for returning to work. But a vote was democratic, and democracy was an important thing.

A clergyman appeared suddenly through a side-door.

Jim supposed that he must have come through the door, but he seemed to have sprung up like a theatre ghost. Jim rose and went to meet the black and white clergyman beneath the crucifix.

# 10

In four houses a coffin lay open on a table. Cheap, stock coffins they were, not made to measure, and that was a difficult thing to bear. It started people thinking whose fault it was, and, of course, it was the Gaffer's. If he hadn't refused to give them a pay-rise, then they wouldn't have been on strike, and they would have been able to afford more expensive coffins for their husbands, fathers, brothers.

The neighbours, the relatives, and the dead men's workmates came in to pay their respects, in whisky and meat sandwiches, paid for by the Union. In one house the coffin was propped upright in a corner, and a full glass put into the corpse's stiff hand. All the bodies were dressed in their best clothes, even if they had to be fetched out of pawn for it. The neighbours helped to buy the clothes back from the pawn-shop.

The mourners talked of the dead men's good qualities, their old jokes, their good and bad deeds, making a story, a saga, out of each anecdote, telling it well and vividly. What with genuine sorrow and whisky, many women began to cry, and then they began to sing, with tears running down their faces into their mouths. They sang 'Soldiers of Christ Arise' like a dirge, which it was in the present circumstances; and they sang the story of the little girl who foresees a pit accident and begs her father not to go to work the next day, but he does, and is killed. The men joined in with their deeper voices, all the people there singing, in the yard at the back, in the house, and outside in the street. They sang of pit and factory accidents, suicides and hauntings, with whisky and sobbing. The children crawled under the table and sat peering out through the fringes of the table-cloth. Some were bewildered by it all, others howled with their mothers. Some, drunk, went to sleep.

A party of young men started up a jaunty, military song about dashing through the lines of the foe, because the atmosphere was becoming too much for them. But their own singing lacked enthusiasm, and when the rest took up the song it became unbearably sad and slow. Men started to cry. One or two sobbed as noisily as the women; others tried to hide in corners, or stood and let the tears run down, refusing to acknowledge them. But they were giving the dead a good send-off. All this crying counted. Years later they could tell how everyone, men and women, had cried at that funeral.

When dark came, candles were lit around the coffins, and a watch was kept by them. Only one or two people watched, because the rest of the family would be needed on other nights, before the burial. The corpses must never be left alone in a room while they were in the houses; nor left in the dark, nor in a locked room. The reasons for this were forgotten, but uneasily close to memory. Still, the watching showed respect for the dead.

\*     \*     \*

Word had been sent round, to Baden Village, to Tividale and Gipsy Lane, that the vote was being held in the 'Tavern'. The men came in.

Methody Bates had been working hard with little bits of paper and coloured crayons. He had drawn a line down the centre of each piece of paper. On the left of the line, in red, he had printed, "GO TO WORK"; on the right side, in green, "STAY OUT". His stiff fingers pulled and pushed the crayon stubs along with difficulty, and his tongue poked from his mouth.

Dewi and Seth explained to the other colliers, most of whom were unable to read, that if they put their cross on the left-hand, red, side, it meant that they were voting to go back to work. If they put their cross on the right-hand, green, side, then they were voting to stay out on strike. The men nodded gravely and crowded around Isaac Walters who had a drawer filled with pencil and crayon stubs, and chalk.

The room was quiet, sun coming in through the door, making deep shadows inside the pub. Men crouched in corners, or bent over their slips at the door, to keep their votes secret in the proper manner. Some went out to the lavatory, where they could lean against the door inside, in privacy, while they decided for red or for green.

Inside the pub Jim Woodall rattled a bucket on the bar again, for the benefit of latecomers. "Votes in here, lads, when thee've done."

Methody Bates finished a batch of voting slips, and drew towards him a sheet of paper ripped from Isaac Walters' account book. On it he had written down the name of every collier who was a member of the White 'Oss Union. He was trying to check that everyone had been in to vote. "Tummy Harrison, Tividale—is he here?"

"He's on his way, Methody."

"Why? Hast seen him?"

"Ar, I saw him comin' on behind we."

"Right, I'll keep an eye peeled for him. What about Ivor Black?"

"He ain't comin', Methody. He's bad. He told me to make his vote for him."

Jim Woodall looked quickly round. "Thee can't do that, Joe. Only Ivor can make his vote."

"I'm his brother, ain't?" Joe demanded indignantly.

"Ar, but thee can't vote for him."

"He told me to," Joe insisted. "I said I would, an' he'll murder me if I don't. He told me to vote for goin' back to work for him, an' I said I would."

"But we can't take it, Joe," Jim explained. "How do we know that's what Ivor wanted thee to say? How do we know thee ain't just gettin' another vote in for thy side? There's only Ivor can make his vote."

"No!" Joe cried. "I'm votin' for stayin' out, but Ivor wanted to vote for goin' back. So I said I'd put it in for him." He came to his feet, holding out two pieces of paper, one with a cross on the left, the other with a cross on the right.

"How did thee get two slips?" Dewi growled.

The men around began to agree with Joe, saying that they'd heard the two men arguing, and that they'd heard Joe agree to make Ivor's vote, and that Ivor's cross was as good as if he'd made it himself.

"It ain't right though," Jim said. "Allow it once, we've got to allow it again. It'd make a mockery of the vote, it would."

"There won't be another vote," said an old man. "First and last time, this."

Jim held out the bucket, and Joe triumphantly dropped the two slips in.

"How did thee get two slips?" Dewi said, but Joe grinned and went outside.

Jek and Shanny were behind the lavatory, sharing a pencil. Shanny was trying to fit the calloused, stiff fingers of his left hand around the little stub. He could not use his right hand because it was bundled up against his chest, along with his right arm. The tip had slipped for a second time while they were digging, and his arm had been broken across a pick by the weight of sliding earth. The local blacksmith had forced the bones painfully back into place, bound the arm tightly with rags, tied the halves of a broomstick to either side of it, and finally made a sling, which could not be taken off until the arm was mended, and nearly choked the wearer every time he moved his head. The arm was cumbersome and it itched a lot.

Shanny got hold of the pencil at last and took it from Jek. With the second and little finger of his free hand he pulled the voting slip from his lips and tried awkwardly to fix it against the wall with the heel of the same hand. "What shall I put, Jek?"

"I can't tell thee that, can I?"

"Well, I don't know what to put."

"Make up thy bloody mind then," Jek said.

"I don't know. It makes no odds now, does it? We've had it now, we have. Thee heard what Jim Woodall said."

"Well, put for goin' back if that's how thee feel."

"Is that what thee'm puttin'?" Shanny asked slyly.

"Does it matter what I'm puttin'?"

"I was just curious like."

"Curious! Like hell thee was curious. I ain't supposed to tell thee what I put."

"Why not?"

" 'Cos I ain't."

"Why not?" Shanny repeated.

" 'Cos I ain't, that's why!" Jek yelled. " 'Cos it's supposed to be so nobody can frighten anybody else into votin' for summat they don't want to. Thee'm supposed to keep it a secret, what thee vote for."

"Well, I ain't about to start bostin' thee up an' makin' thee vote like me, am I? Not with me arm tied round me ear-'ole."

"Nobody could make me change my mind anyroad," Jek said.

"Oh hark at him," Shanny said sarcastically.

K                                     145

Shanny turned back to the slip, pressed against the wall by his wrist. He tried to put the pencil against it and dropped the pencil. Hissing, Jek picked it up and put it back into his hand.

"Shall I put for goin' back to work then?"

"If thee like," Jek said flatly. Shanny squinted sideways at him and asked, "I'll put for stoppin' out then?"

"Put whatever thee like, but for God's sake hurry up!" Jek shouted.

Shanny squirmed his hand against the wall and the paper. He was about to make the cross when he stopped. "Jek—which is which?"

"Eh?"

"What do the colours stand for again?"

"Oh bloody hell. Green for stay out, red for go back to work."

Shanny, tongue between teeth, made a staggering 'X' under the green letters. "There—that's what thee'm goin' to put, ain't?"

Jek snatched the pencil. "It's no business o' thine what I put." So Shanny kept peering over his shoulder and ducking under his arm as he made to draw, pushing him aside and repeating over and over again, "That's what thee'm goin' to put, ain't it?" in an imbecile voice, his face pulled to match. "Oh, clear off, bloody Shanny!" Jek shouted, and hit out in exasperation, catching Shanny on his broken arm. Shanny stepped back, quiet. "I only wanted to know what thee put, I don't see as it can hurt anythin'."

Jek was afraid that he had hurt Shanny badly, and he hadn't meant to. He scrawled a cross under the green letters and folded the slip twice. "Well, I ain't tellin' thee!" He turned his back and walked towards the pub to hand in his vote.

'STAY OUT.' He wasn't sure in his own mind that he'd

made the right vote. But he had stubbornly ignored all doubts and had voted for staying out anyway. He even hoped that enough of the others would do the same to keep the strike going—but that was stupid, they were finished. Even with the strike pay and odd jobs they were going hungry for two or three days, and children were sick and dying. Without the strike pay how would they manage? And Jim Woodall had made it clear that if the strike went on, it would be without strike pay.

He felt as if his guts were itching and writhing about inside him. He would hug Rachel's words to him like a blanket to hide from ghosts under—you didn't have to think of the people, only 'The Cause'.

Then he would see Nellie's face, white under her dark hair, heavy-eyed and ill. And the woman who had married a strike-breaker and had to suffer being pushed and kicked. His mother, too, worried to death, irritable—he couldn't find much love in himself for Reenee, but he understood that she'd had a hard life, and he pitied her.

Oh for God's sake! Why couldn't he have half a mind, like black and white, and see only half of everything? His half, the half he wanted to see. Why did he have to see a thing from all directions, even a blackleg's point of view? It only made life more difficult and life was difficult enough already.

Why couldn't he hold to one opinion for more than a minute without doubts coming in and tumbling it down? He held the vote tightly in his hand and ran to the pub door before he could rip it up.

\*     \*     \*

"I hope thee'm satisfied," Reenee said. "Four good men and a couple of babbies dead (but I don't suppose they count, do they?) and Our Nellie bad. I only thank God that the Gaffer's such a gentleman that he ain't had thee all locked up, nor throwed we out the houses."

"I told thee as only trouble'd come on it," Grandad Ellis said wisely. "Thee can't alter the way thee've been made."

Jek glanced at them both sourly, and half wished that Dewi was there. He'd sort 'em out. "Mother—is there anythin' else to ate?" Aynoch asked, chewing the last mouthful of a slice of bread. His stomach rumbled.

"No, there ain't!" Reenee snapped. "An' thee'd better make the best of that bread, my lad, if thee ain't back at work soon. There's goin' to be another mouth to feed."

They all looked at her. "Another babby, Mother?" Jek asked.

"Ar," Reenee said. "It'll be called Sam if it's a chap, an' Mariah if it's a wench."

"What'll it be, thee reckon, Muth?" Jek asked.

"I reckon a chap meself, an' I've only been wrong once, an' that was over Liam Shannon, not one of me own."

They sat, taking this in, and adjusting to it. "Hey—our father'll be as proud as punch," Jek said. "Shall I go an' find him an' tell him?"

"Thee stop where thee am!" Reenee yelled. "I ain't tellin' him until I got to. He'll be savage. What we goin' to do for money?"

148

# 11

The four coffins lay in the cold church. On each box lay a
large wreath, with a card expressing the Gaffer's regret.
There had been sharp disagreements about the wreaths.
Most of the women, some of the men, felt it was a kind
thought, and proved that the Gaffer wasn't as black as he
was painted. Others, Dewi among them, considered the
wreaths to be an insult; that the Gaffer was laughing up his
sleeve at them. To Dewi's mind any gift from a Gaffer
could only mean something of this sort. Jek wasn't sure.
Like his father, he felt insulted by the shiny, obviously
expensive wreaths; but he thought that the Gaffer was only
a man, and might feel genuinely sorry that four of his
employees had been killed, even if they'd been strikers.

Anyhow, no one could help but notice that the wreaths
completely outshone the bunches of wild flowers provided
by the colliers and their families. The coffins were
covered with late dog-roses, moon-daisies, yellow shoes-
and-slippers, flax, ragwort, honeysuckle and bindweed;
but these flowers only made a background the better to set
off the wreaths. It had been no easy job to find all those
flowers around the factories, forges, marl-holes and
quarries, either.

The colliers crowded into the pews with their families;
strikers and blacklegs, although neither faction spoke to
the other. Those who could not find a seat stood packed

at the door, or in the aisle. The cold and silence of the church was broken by constant coughing and hawking, and feet shuffling.

The clergyman, all black and white, stood high in his pulpit. He filled the church with his buttery voice.

There was a prayer for the dead, the widows and their families. It droned on and on. Most of the congregation, not knowing the words, made slurring noises, or spoke only the last half of the words. It was all the same to them—they were giving the dead a good send-off.

Now the clergyman began a sermon. "A father loves his son, we all know this, don't we? Or we know instances of it. For example, we have Abraham's love for Isaac, even David's love for Absalom, for love endures through differences and quarrels. But we all know too, do we not, my friends, that sometimes a father, however loving, must punish his son. If that son does wrong, then he must be shown that he has done wrong, and be brought back once more to the path of righteousness. This does not mean that the father loves the son any less, for the punishment is given with a heart full of love, is given, indeed, *because* the father loves the son."

Jek, sitting between Shanny and Ayli Black, thought of Dewi drunken or angry, and grinned ironically at a wall carving of a bearded man with leaves shooting from his mouth.

The Vicar smiled at them all, and explained kindly, "When we speak of a father, we do not always mean our earthly father. Sometimes we call God our Father, our Heavenly Father, because He is the Father of everything that exists on earth. And like a father, He must sometimes punish His children when they misbehave, to bring them back to the straight and narrow path."

Jim Woodall suddenly saw the way this sermon was going. He turned his head to see Dewi beside him, and

Seth. He knew by Dewi's folded face, bottom lip almost touching his big nose, and by quiet Seth's scowl, that they understood too.

"The Bible contains many instances of this punishment, my friends: Adam and Eve being driven from the Garden, the destruction of Sodom and Gomorrah, the Flood. There are more evidences of God's punishment and love that are not written in the Good Book: the Great Plague, of which you will all have heard, was sent to punish our ancestors for their sins; and certainly, the terrible visitation of Cholera which came upon us ten years ago was sent from God to punish the dreadful drunkenness and licentiousness which was so prevalent in this district. (And still is very prevalent, my friends, and I tell you plainly that if this swinish guzzling in public and private houses does not cease, then the Lord will scourge you once more with the Cholera, the Fever, the Flux, or the Influenza—)" The Vicar broke off and coughed. Those close to the pulpit swore that he muttered, "Where was I?"

"We are gathered here today, my friends, united in our grief for our dead brothers, sadly cut down in the prime of their lives. We know that they are gone to a better, happier life, but it is hard for us because we selfishly loved them and we wish that they could still be with us."

The clergyman's voice had been growing more and more buttery until Jek felt like retching. His gaze wandered about the church, picking out the faces of the colliers he knew, strikers and blacklegs. Many of the faces were angry or bitter. Jek didn't understand.

"Yet, even in our sorrow, we must ask ourselves: *Why* did God take these, our brothers, from us? The answer must be, my friends, that this is a punishment upon us for some wrong we have done. A gentle slap on the wrist, as it were, compared to the spanking that was the Cholera epidemic."

151

A woman began to cry. Jek felt his face flush. Four good colliers a 'slap on the wrist'?

"The next question we must ask ourselves, my friends, bearing in mind that God loves us always and is cruel only to be kind, is: What wrong have we done to deserve this punishment?"

Jek understood. He looked sharply at Shanny who turned to him at the same time. But Shanny's face was blank. He didn't see what the clergyman was going to say.

"I think we know, in our hearts, what we have done wrong. I know, my friends, that you will not blame me for speaking frankly in this House of God: you have disobeyed your Masters, and, in doing so, have disobeyed your God."

The black scowl on Dewi's face was very unsuitable for a church. Shanny, light suddenly breaking on him, glanced at Jek, and for a second thought that his uncle was beside him, not his cousin.

"My friends, as surely as any of you would punish an erring son for disobeying the rules you made for his well-being, so will our Heavenly Father punish this breaking of His laws if we do not return to His ways. If this wicked strike—"

He had said it! The church began to bubble like a boiling pot.

The clergyman raised his greasy voice above it. "*If* this wicked strike does not end soon, we can look to see another visitation of the Cholera, which . . ."

Dewi rose from his seat, his clogs clashing on the stone floor. He pushed his way to the aisle and snatched the wreaths sent by the Gaffer from the nearest coffins. With the wreaths slung on his arm he touched his brow to the coffins, turned his back on the pulpit and marched out. Close behind him was the eldest son of one of the dead colliers, carrying the wreath from his father's coffin.

Jek said, "Me Dad," and was up, struggling along the pew amongst the feet of the others. He ran down the aisle to reach the two men, and picked up the last wreath. He too raised his hand to his brow as he turned from the coffins. The three of them walked from the church, pushing through the men packed at the door.

Jim Woodall stood up. It seemed disrespectful to leave the coffins, and walk out; but wasn't it more so to stay and acknowledge this Gaffer's sermon? Jim and Seth stood up, and every man in their row followed them. They filed past the coffins, raising their hands, and left the church. One by one the other strikers came from their seats, and from the crowd at the back, saluted the coffins while ignoring the clergyman, turned their backs on pulpit and altar, and walked from the church. Even some blacklegs left.

But most of the blacklegs stayed behind and heard the sermon to the end. Because they agreed with the clergyman's words. Or because they felt that they had parted from the strikers and had no right to join them outside. Or because they didn't know what to do, and so sat, or stood, frozen into place. They heard the Vicar, when he had exhausted his threats of divine revenge, read messages of sympathy from various local and well-to-do personages— including the Gaffer.

At the end they filed out to see the Gaffer's wreaths thrown into the mud of the road, carts passing over them. And, on the other side of the road, the strikers waiting for the sermon to end, so that the families of the dead men could return to the church, to watch over the coffins until their burial the next day.

Walking back home was a miserable affair. It took the church to rub holy salt into the wound.

Jek straggled at the back. During the wait outside the church he had had a lot to think about. The cholera, for instance. He had been only eight at the time of the epidemic, and it had touched Oldbury only lightly. But the memory of it, his parents' fear that it was coming nearer, their talk of it, the streets it had emptied and the graveyards it had filled was like a black fog hanging in the dark places of his mind. For years afterwards he had thought that everyone who died had died of cholera.

This strike—what good had it done? The newspaper had been right. And here was the Church of England, preaching what his grandfather had preached months ago: that no good would come of the strike because it was against God. But the Church promised something worse than 'no good'—cholera.

Jek dropped further and further behind the other colliers. He turned on his heels and ran back up the street, between two houses and into the next street, tumbling muck-heaps as he went. He twisted, walked, ran through the alleys and gulleys and refuse to the fields; and then he settled to a hard, fast walk.

He was nervous, uncertain, frightened. What if the strike brought the cholera back to top off all the trouble it had already caused? And he had supported the strike, he would be partly to blame. If there was a God, if the preachers were right. If. Why couldn't he stick to one opinion for five minutes together? Why must his mind toss from one side to another until it felt as if his brains were boiling out of his skull?

It was a sensation he was familiar with. But the last time, Rachel had given him an argument to stop the arguments inside him. It was like a poultice. When the doubts and the brain-boiling began, he produced Rachel's argument and thought on. If he tried hard enough he could convince himself that it was true. He was going to Rachel again.

He ran up the muddy Dudley street, and the children recognized him, and came shouting, "Them from Oldbury ain't no good, chop 'em up for firewood!" He dodged away from them around the corner of the cottage.

There was the strip of yard between the forge and the cottage door, cluttered with rusty iron and bad nails. Thrown against the forge walls was their shovel and the pick. Hens pecked over the muck-heap.

Jek poked his head in at the forge door, because yellow light came from it. There would be hammering too, but that was lost in the general thunder of hammers all around. The Ansel family were inside, crowded around the fire, working fast, risking hands and fingers at every move, either under the hammers, or in the fire. Sparks flew, bars of white-hot metal were swung above unprotected heads. A careless movement, or a stumble on the crowded floor could have had someone in the fire, or branded. The din blurred Jek's mind. He couldn't make himself heard above it. His mouth opened and closed silently, like a fish.

One of the women saw him and pointed. Eliakim Ansel stopped work and swung around. Everyone else stopped and stared at him vacantly.

"It's thee," Eliakim said. "Where thee been? Ain't seen hide nor hair of thee or thy mate. Where's my coal, eh? Where's my coal I was promised?"

"We ain't got no coal, an' we can't get thee any."

"Oh ar! So much for promises. Good job I didn't give thee any money, ain't? What thee come round here for then? I ain't got anythin' for thee."

"I didn't come for anythin' of thee," Jek said, "except we tools back. I come to see Rachel. If I ask her to go with me for a walk, thee won't mind, will thee?" Jek made the request as if it was an order.

But Eliakim grinned broadly. "Aah. I thought thee was interested in that quarter. I've heard about thee visitin' late at night behind me back. Thee take her, an' mind thee bring her back."

Jek nodded, unsmiling, and made to leave the forge. Eliakim called him back. "What happened to my coal?"

Jek stared at him coldly for a second. He said, "What d' thee want to know for? Thee'm workin', ain't?"

"For the first time since thee left, my lad. I managed to buy a bit o' coke from a man I know. I managed to scrape the money together. I'd be all right if I was still waitin' for thee to bring that coal thee promised. We got to live as well as thy blasted colliers, thee know. What happened?"

Jek said, "The day we was goin' to get thy bloody coal there was an accident. There was four men killed. The Union can't pay for the strike an' their funerals, so the strike's finished. We had a vote, an' we'm goin' back to work. If the Gaffer'll let we."

"I heard about the accident. It was thy pit, was it?" Eliakim asked.

"It was," Jek said. He turned his back on the forge and crossed the yard to the door of the cottage.

"Oi—Welshy."

Jek turned. Eliakim was at the door of the forge, holding a bar of white-hot iron in a pair of tongs. "Thee watch thy tone o' voice when thee speak to me—an' thee mind as our Rachel don't get hurt, else I'll mark thee, my lad, I'll mark thee." He jerked the iron bar and the burning flakes of scale from it flashed across the yard.

Jek watched the bright specks settle. Then he said, "Just like bloody Hamlet. Go cut thysen, Ansel, thee don't scare me." He went into the house. Rachel was sweeping up.

"Oh. Oh," she said. "Oh, it's thee." She hadn't heard anything of the talk outside for the clamour of the forges. As far as she was concerned Jek, absent but not forgotten, had suddenly materialized before her broom-head.

Jek hunched his shoulders and stood on one foot, kicking

156

the floor with the other. "I wondered if thee'd like to come for a walk."

"Oh," Rachel said. "I'd have to ask me—"

"I asked him."

"Did he say I could?"

"Ar! What does it matter? Am thee comin' or ain't thee?"

"Thee'm in a nice temper," Rachel said.

"Am thee bloody, pesterin' comin' or ain't thee?"

"I'll fetch me shawl," she said coolly, and went into the bedroom. She was back in a second, an old blue shawl about her head and shoulders, and Jek grinned at her. He followed her out into the yard. Work was going on again in the forge, and there was no sign of Eliakim, or his branding iron.

They went up the street side by side, neither speaking. Then Jek said, "There was four men killed. They was diggin' in the tip for coal to sell an' it slipped."

"Oh what a shame," Rachel said.

"Is that all thee can say?"

"What else *can* I say?"

Jek sulked for a second or two. "It was 'cos of the strike. If it hadn't been for the strike they wouldn't have been diggin' in the tip an' they wouldn't have been killed."

"Now what am thee on about?" Rachel demanded, her shyness wearing off.

"Well, me Grandad said right from the start as the strike was against God an' it'd only bring trouble. An' it has brung trouble. To everybody, just like paper you read for me said. An' today, at the funeral, the Vicar said as God killed them men as a punishment, an' that he'd probably send the cholera again an'—I don't—properly believe in God, but I keep thinkin'—what if there is one an' what if He does—send the cholera, I mean—? Then we'd be to blame an'. . ."

Rachel was shaking her head in astonishment. "Oh, thee am daft. Honestly, I've never knowed anybody like thee afore. I've never knowed anybody so daft. Thee'll believe anybody, won't thee? Look. All the pit accidents there is every year—there's dozens, ain't there?"

"Ar," Jek agreed eagerly.

"Well, is everyone o' them a punishment from God? They're just accidents except when some Gaffers or the Church want to frighten thee. Thee know that!"

Jek watched his feet and said nothing.

"Why do they pick on this accident and say it's a punishment from God? To frighten thee, so thee'll never be naughty an' strike again, that's why. The Church is always hand-in-glove with the Gaffers, 'cos the Gaffers am Church of England, an' the Vicars know which side their bread's buttered on. Now thee know that an' all."

"Thee'm right!" Jek said happily.

" 'Course I'm right. Now *thee* tell *me* why this accident happened. Thee know more about it than I do—thee tell me."

"Well," Jek began, "they was cuttin' their galleries narrow, an' they wasn't proppin' 'em right. Tryin' to do it fast, see. That's what Shanny ses. That's what we'd have been doin'. We was lucky, it could have been we."

"There thee am then. Punishment from God!"

"Thee'm right," he said. He slid her a look from the corners of his eyes. "Thee'm clever —— Shall we go to the Castle?"

She smiled and nodded.

They turned into Market Street and walked along to Castle Hill. As they walked, he told her about digging for the buried men, the vote, and the service. She stepped beside him, arms folded under her blue shawl, and said, "Ooh. Aah. How horrible. Sooner thee than me. What a shame. Aah." But she means what she's saying, Jek thought, and his spirits rose.

They went in at the gateway in the Castle's old outer wall, and climbed the steep path inside that led up to the Castle itself. The bridge across the dried, deep moat had partly fallen, but the low parapet on one side remained. They could run along that, Jek said.

"Oh no. We might fall. It's ever so far down into that ditch."

"It's all right," Jek assured her. "Look—it's wide enough for two carts to pass one another."

"It might just be wide enough to put thy foot on," Rachel corrected.

"It's all right," Jek repeated. "We always go across it." To demonstrate he stepped out neatly along the parapet in his clumsy wooden shoes, reached the other side and came half-way back. "See? Come on, it's all right. Here's me hand, I'll help thee."

"No, I don't like it. Let's go round to the other gate."

Jek clicked his tongue. "We don't want to walk all the way round. Come on. Thee won't fall."

"I'm scared."

Jek grinned, and felt twice his natural size and height. He wasn't scared. "Thee won't fall! I'll hold thee hand."

"Thee holdin' me hand won't stop me fallin', an' if I do, I'll take thee with me," Rachel said.

"No—I'd let go," Jek answered her gallantly. "Go on then, walk all the way round if thee want to. I'm goin' in here." He turned and went clattering towards the castle along the parapet. Suddenly he tottered, one leg out, arms flailing.

"Aow!" Rachel shouted.

Jek righted himself, and hopped round to face her, balancing on one leg. "Did thee think I was goin' to fall?" He leaned out over the drop, on one leg.

"Don't do that," Rachel begged, "please."

Jek stood on two feet again and came back to her. "Come on. We can go slow an' careful. I bet I could hold thee anyroad, if thee was to fall. I'm strong, I am. I can fill twenty tubs a day if I try, like the old uns."

Reluctantly Rachel said, "All right then—but if I get scared, promise I can go back?"

"I promise," Jek said.

She gave him her left hand, the good one, and he towed her slowly a foot or so along the parapet. She watched every step her feet made, but tried not to look past them into the drop. Then Jek began to walk faster, pulling her along. "No," she said, "stop it," and tried to drag her hand free, but he held tight and began to run. She had to run too, terrified that at any second she was going to drop over the edge and down into the deep moat.

Jek pulled her off at the end of the bridge and immediately let go of her hand. She was gasping and giggling, frightened and excited at once. "Thee promised," she said, "thee promised."

Jek only grinned. They went into the castle, passing through the long, roofed gate where the portcullis had hung, and the draw-bridge had come down. Their footsteps echoed, and their feet splashed in puddles among the cobble-stones, because it was always damp and mossy in the gate-house. Jek showed her the stone of the gate's arch where his Grandfather Davies had carved a pick and a man, being unable to write his name.

The courtyard, when they reached it, was overgrown. Tall grass, green and yellow, purple marl-hole flowers, nettles, brambles, even small trees grew up thickly all around, making a rich green smell that—almost—defeated the smell of chemicals and forges. The towers of the castle rose high above them, grey limestone grown with yellow lichen; hollow towers with great square windows framing the smoky sky behind, bushy clumps of red and white flowers growing where the window-seats had been in upstairs rooms. There were no other people there, no birds; the place was silent, and damp, and depressing. It reminded Jek of the funeral he had been forgetting and the grin left his face.

"Let's go up the Keep," he said, suddenly turning to her.

"No, it's dark in that Keep."

"It's only dark in one little bit. Thee go up the steps, then it's dark at the corner, then it's light again. The dark only lasts for a second if thee carry on instead o' bein' frightened."

"Ar, an' thee said as thee'd let me go back if I got scared on that bridge," Rachel said sceptically.

"I'm tellin' truth this time, honest. Come an' see. Thee can see everywhere from top of Keep. Thee'll be able to see thy house an' mine."

"Oh—all right," Rachel said.

The Keep was solid and square, the steps narrow, steep and uneven. Holding Jek's hand—she was careful to give him her left again—she scrambled up behind him. There came the corner where it was pitch-black, like a cave, never lit. She hadn't climbed higher than this before, because she had always been frightened. She hesitated now, but Jek pulled at her hand and she went on. He had been telling the truth this time. After two or three steps the dark vanished. But up and up, round and round, the steps continued. They were both panting when they reached the top.

What a view. What a sprawl of black roof-tops, smoking factories, sulphurous forges, canals and grimy railways, black sidings, pit-banks, marl-holes and quarries; chimneys, sheds, pubs, holes in the ground, everything black and blackened; the countryside blackened, the Black Country. A haze of smoke and dirt rose up from the huddle and hung protectively over it all. Protection from the sun. Most things that grew here were stunted or lank, colourless because the sun could not reach them through the smoke.

Jek turned round and round to see it all, and the view filled him with a kind of happiness. He knew that the people who lived in those houses were exploited, mostly miserable; but he reckoned that his people wouldn't get a better deal in the country. And this smoke and dirt he was used to, he knew it, it wasn't strange like the countryside when he went poaching with Shanny. He enjoyed seeing so much of it spread out below him. He leaned his elbows on the wall and stared out. "They reckon thee can see Wales from here—but I wouldn't know it if I did see it. Got a lot of mountains, I've been told, Wales has, an' castles like this, an' streams an' lakes, big pools like. Me Grandad Davies' father come from Wales. From Merioneth, so they reckon, wherever that is. I know the name of the place he come from—I've been told it—but I can't say it right. Wynne Davies me Grandad's father's name was. He come here 'cos he reckoned he was bein' put on an' done down where he was. Cheated. An' he had his *own farm*!" he added.

Rachel nodded. "What's thy name?" she asked.

"I thought thee knowed!"

"No," Rachel said. "Thee never told me. Thee tell me tha Grandad's name, but not thee own."

"Jek Davies it is—Jek's short for Jechonias."

"Oh," Rachel said, and nodded again.

Jek left the wall and slid nearer to her. "Our house'll be over this side." He studied the complicated landscape. "There's thine. See it?"

Rachel looked in the direction he pointed, bobbed her head about. "Where?"

"*There*. See it?"

"No—oh ar! I see it!"

162

Jek slid nearer still. "An' our house'll be over there somewhere." He squinted. "See them pit-banks?"

"No. *Which* pit-banks?"

"Them there. They'm right there, thee must be able to see 'em."

"Well, I can't."

Jek, tongue nervously between teeth, made this an excuse to put his arm around her shoulders. He pulled her awkwardly to him. "There, look, there," he said, with a dry mouth, his voice rising a little. "Thee must see it."

"Oh ar, I see it now," Rachel said quickly. She had become very stiff; even stiffer than Jek himself. They stood there for a minute, like peg-dolls, glaring fixedly towards Oldbury. Then Jek, in for a penny, in for a pound, ducked his head to her face, like a chicken pecking, and succeeded in bumping the end of her nose with his lips. Shocked by his own daring he took his arm from her shoulders and went twisting quickly down the Keep steps, leaving Rachel to come down by herself, even through the dark corner.

He waited for her in the courtyard, but only to see her come out of the Keep door; then he turned his back on her and hurried to the gate. But Rachel wasn't going to run along the bridge parapet again, certainly not by herself, so she left by the other gate, where the bridge was in good repair. She half ran, half walked around the Castle, to see Jek waiting by the gate. But when she was still some yards from him, he turned away from her again, and went down the path in front of her.

A little angry, more flustered, Rachel scurried after him, pulling and holding the blue shawl around her. Then it occurred to her that he was trying to get away from her, so she said, "Damn him!" and slowed to a walk, looking carelessly about at the scenery, as if she had come out for a stroll alone.

Jek came back up the path to her, red-faced and

grinning. He held out his hand for hers. Craftily, she began to circle around him, so that she could give her left hand, but Jek moved with her, and he moved faster. He took her thumbless, stump-fingered hand in his own and held it tight.

They went all the way down the hill to the Castle gate, and along Market Street, and through all the forge-lined streets without saying a word, because neither could think of anything to say. Jek's throat was as dry as an old bone. They didn't even look at each other; Rachel stared at the walls by her side, at all the bricks neatly dove-tailing into one another; Jek stared vacantly across the road to the walls on that side. But between them, in the folds of Rachel's wide brown skirt, their sweaty hands were joined.

The children near Rachel's home recognized the foreigner from Oldbury again and chased after them at a safe distance, shouting, "Rachel's courtin', Rachel's courtin'! Them from Oldbury bain't no good, chop 'em up for fire-wood, an' when the fire begins to crack, the fleas'll all run down their backs!"

At the Ansels' cottage Jek ducked his head and gave her another peck on the cheek, and then made off down the road, heels high, without a word, or a wave, or a backward glance.

Rachel wasn't worried. She understood him now.

Jek crossed the fields, brooding on the sinfulness of his own character. Luring innocent wenches to Castle Keeps and there putting his arm around them and—and kissing them. And after a funeral too. He suddenly remembered the tools, the pick and shovel, and came to a dead stop. He looked uncertainly over his shoulder. Would he go back for them or not?

Then he grinned and swung on towards home. Leave them. Leave them for as long as he could. The longer they lay in the Ansels' yard, the more times he could use them as an excuse for visiting Rachel.

He found a stick, and began to swing it from side to side across the path, lopping the heads off nettles, and plantations of scented marl-hole flowers that buzzed dangerously with wasps and bees. After all, the funeral had been that morning. You couldn't go on being sorry for the dead for ever.

# 12

Someone had to tell the Gaffer that the colliers were ready to return to work. It couldn't be Jim Woodall, because the Gaffer had sacked him, and to send him as their representative would hardly be tactful.

So they asked Dewi. It was not an errand he liked. The thought of oozing and scraping round the Gaffer made all the hairs at the back of his neck swivel up on end. It made him feel sick and he knew he couldn't do it. He would only make the situation worse. He offered the job to Seth Jones.

Seth, everyone felt, was the ideal choice. Why hadn't they thought of him before? Seth was always soft-spoken, turning away trouble if he could. He had as much pride as any other man, but knew when to forget it. Seth would have no trouble in controlling his temper.

Seth left early in the morning, and the colliers settled on the door-steps, on the pig-sty walls, on the pub-steps, to wait.

Reenee was ironing on half of the table, banging away with a heavy iron which she heated against the bars of the grate, by fitting it into a little stand which hooked on to the front of the fire. Her sharp elbows whisked backwards and forwards as she worked, and her tongue went faster as she cursed all her children and the one to come. Those that could left her to grumble to herself.

Jek was half-way to Shanny's when he decided that he

didn't want Shanny's company. He didn't want anybody's company. Except Rachel's, but she was in Dudley.

He wandered over the fields, around the edge of the quarry. He looked down its steep, corrugated sides into a pool of black water, then drifted on aimlessly. End. End of the strike and end of living. For sixteen weeks he'd had the chance to sleep, idle and work as and when he'd wished. He'd had choice. Well—not really. He'd had to take odd jobs, for money. But it had felt like choice, it had seemed like it. He'd never have that again.

So why walk? He wasn't going anywhere. He sat down in the long grass, then lay down, arms beneath head, legs stretched straight out. Worn clogs; shabby, baggy, thick-seamed trousers; cheap cotton shirt, thin as tissue paper across his shoulders; shapeless, lumpy old jacket and old, limp cap, colourless with age and dirt. He stared up at the sky, which could be seen as a bluish tint beyond the smoke.

He felt empty. After the end there is nothing, so he felt nothing. Numb and empty. There was a heaviness in him but it was not an emotion; it couldn't be defined and called fear, or disappointment, or sadness, or anything that he should be feeling.

He stared up at the staring sky above him and began to forget where he ended and the sky started. He didn't know whether he was on his back, or his side, or his head; or even on the ground. He dimly imagined himself to be spinning, like meat on a spit, but in all directions, heels over head, round and round. Sixteen weeks of escape, and now life was clamping down on him again. Up at half past three each morning, to be down the Pit by six—a two-mile walk to the Pit, then a long, long crawl underground to the face. But you were paid only for the coal you dug, not for the travelling, so you had to make sure the crawling was done in your own time. Then hard graft in the close tunnels underground, desperate hard work, to fill enough tubs in the time you had, to earn something worth calling a wage, remembering that any stone in the tubs lost you the payment for that tub. Working stark-naked, with sweat pouring, streaming from you, because it's hot underground. Sweat wasn't the only wetness though; the pits all ran with water, underground streams and pools. It made the work more uncomfortable and dangerous.

After fourteen hours of back-breaking work, sometimes longer, the collier had to crawl and stoop back to the Pit-head. Then a two-mile tramp across the fields if you lived at White 'Oss Row, eat the dinner Reenee would be keeping warm for you, and collapse into bed, to be up again at half past three, quarter to four the very latest, and out of the house for another fourteen hours' work—day in, day out, six days a week.

On the seventh day, the day of rest, you rested. Usually you were too tired to do much else, sleeping well into the morning. Then you crawled out of bed for a look at the dreary old world, with the thought of work the next day hanging over you. But this was the only free day you had. You had to spend it in drinking, courting, visiting, cockfighting, whippet-racing, poaching . . . you had to get up and stop being tired.

If you were lucky and could win a bet, you could have a day or two off work, live on your winnings; if you were lucky. But few people were lucky, there was the rub. If they were lucky they were rich; and if you were rich—who needed luck? Otherwise it was work, work, work, working yourself away for a bare living, all your life.

He couldn't stand to see the empty sky above him any longer. He rolled on to his belly and hid his eyes in the warm, dark crook of his elbows. The warmth and the dark reminded him of Rachel, but Rachel was in Dudley. She couldn't give any protection against the emptiness; not from there, anyway.

End of the strike, the one high point in his life. Only see Rachel on Sundays. A walk to Dudley to see her, walk back, and up at half past three the next morning. Oh Christ, what a life. A black and narrowing tunnel, no light, just on and on, blacker and narrower. Bleak and empty. But short; with luck. Rachel, I wish you was here with me.

\*　　\*　　\*

Seth came back. The Gaffer had been very polite. He sent his condolences to the new widows in their tragic loss. The colliers would return to work tomorrow, and the whole matter would be forgotten. But they would have to work longer hours than usual for an indefinite time, until losses were made up. Also, he regretted that he must inform them of a wage cut—their wages would be cut from a penny a tub to three-farthings a tub. He further regretted that a wage cut to a half-penny a tub might prove necessary. Owing to the strike, the White 'Oss had made quite considerable losses. It was his duty, as manager, to make up these losses to the owner. He understood that the reductions might cause inconvenience, even hardship, to the colliers, but they must bear in mind that they had brought it on their own heads.

Under no circumstances would the Gaffer employ the man Woodall again. They would understand his point of view; the man was a troublemaker. He appreciated their concern for their work-mate, but found it impossible to grant their request.

So Seth repeated the Gaffer's words, which he remembered very well, unrolling layer after layer of them. The colliers listened in silence. They were in no position to make any comment. They were beaten.

Many of them felt relief that the strike was over. They didn't have to fight any more, search for odd jobs, meet

169

with dislike everywhere because they were making trade difficult for everyone. Now they could sink back into the old way of life, which they knew. It was hard, the new wages would make it harder, but it was safe and worn, like an old shoe. No effort. Shanny even grinned. They'd tried, they'd lost—so forget the whole affair. They'd always managed, they would manage now. They'd live.

But Jek didn't have a forgiving nature. He remembered. All the Gaffer's words, the nailers' swearing, doors shut in his face; he remembered. All the shame of having to crawl to the Gaffer and go back to work for a lower wage, after saying that they would win the strike. All the added bitterness of knowing they had brought it on their own heads, the wage that might cause inconvenience, even hardship. These things lodged inside him like thorns, chafing and stabbing and scratching. Jek wouldn't forget or forgive.

\*     \*     \*

The Gaffer wouldn't employ Jim Woodall. No more would any other Gaffer. The word had obviously been passed with the port that James Woodall, ex-collier, of the Green, Oldbury, was a troublemaker, and an agitator. As soon as he gave his name he was turned away, or asked, "Is that Woodall from the White 'Oss?" And when he said, "Ar," he was shown the door. It had occurred to him to give a false name, but at the last minute he always changed his mind. He was James Woodall, Union Man, and why the hell should he deny it?

So there was nothing to be done except leave the district entirely, and go somewhere else, where the Gaffers had never heard of him.

He left, with his wife and children, in the early morning drizzle. The colliers saw them off. Thomas Shannon gave

Anne a poached rabbit, and Dewi presented Jim with a clatter of copper coins, collected from the strikers of the White 'Oss. Jim said thanks; Dewi said thanks from all of them; then Jim sighed and grinned weakly, hoisted his heavy bundle and trudged away down the muddy street with his family. The colliers stood very still and watched them go. Not an emotional farewell; just a wet one.

As the Woodall family sloshed along, they were joined by a man who came out from a dripping alley-way. It was Jek. He walked with them a few paces, then said, "I'm sorry thee've got to go, Jim."

"An' I am," Jim said, attempting humour.

"Where thee goin' then?"

"I thought we might go to London," Jim said.

Jek thought about this, with a glance at Anne. "It's a long way."

"I might get work on the way there. Thee never know".

Jek squelched awkwardly on beside them, wondering what to say. He thought they'd all catch pneumonia before the end of the street. "I am sorry, Jim, honest—Annie, I am."

"We know thee am, love," Anne said.

"It ain't the end of the world," Jim said, "is it? We're alive, it could be worse. We'll make out in the finish, thee'll see—I might write thee a letter, tell thee how we get on."

"Ar—do that!" Jek said, knowing as he spoke that Jim would never write. They'd never know how Jim—and Anne—got on. "Well—tara," he said, "good luck." He held out his hand and Jim shook it. Then Jek turned and walked back towards the Row, the rain trickling down the back of his neck. When he turned again, walking backwards for a while, there was only the long, blackened muddy street, hammered on by the brown rain. The Woodalls had turned off somewhere.

Well, they were gone. That was the end of them. He

hoped that they would manage well enough wherever they were, but he couldn't imagine them existing anywhere else but here, the Row, the Green and the Pit. The world was a very small place. It began in Oldbury and ended in Dudley. The Universe revolved around the White 'Oss Pit. Beyond Dudley there was void; if you went too far down Rood End in Oldbury you would drop off the edge of the world.

Jim Woodall and his family had passed into another world altogether; already they were ghosts in Jek's mind. Jim wasn't Jim Woodall any longer, with a whining voice and too much readiness to believe other people. He was the Union Man. Anne wasn't Anne Woodall, she was the Beautiful, Brave Union Man's Wife. Driven out of their home by the Wicked Gaffer. Jek wouldn't forget that, and he wouldn't let anyone else forget it either.

\*　　\*　　\*

It was quarter past four in the morning. Still dark. And cold and wet. The colliers were trudging to work at the White 'Oss Pit across the fields. They straggled into two groups: ex-strikers and ex-blacklegs. The men who had worked together in gangs at the coal-face had re-shuffled, so that the gangs were now either all strikers or all blacklegs.

When the blacklegs had heard of the accident, they had run to the tip and had slaved as hard as anyone. But now they called themselves stupid for doing it. They had no reason for helping the strikers, nothing to thank them for. The strikers had been selfish, not listening to sense; now all the wages were cut, the blacklegs' wages too. It was the strikers' fault; and they hated the strikers for tricking them into digging at the tip.

The strikers wouldn't relent. The blacklegs were the Gaffer's lick-spittles, like the foremen, and no sucker-up to

172

the Gaffers could be a friend of theirs. They hated the blacklegs for their condescension in digging at the tip.

It was stupid, Jek thought, but he couldn't escape from the same feelings himself. So there it was. He hated the blacklegs.

But now he had to work through the next days with as little thought and as little feeling as possible, until Sunday. Then he could see Rachel.

And on Monday he would submerge into work and dim existence again—until Sunday. Then he could see Rachel.

And on Monday. . . .